TISSUE INTERACTIONS
AND DEVELOPMENT

Also available in this series: *Elements of Human Genetics* by L.L. Cavalli-Sforza, Stanford University School of Medicine.

About the Cover:

The micrograph on the cover of the soft-bound edition of this book is of a chick embryo at the 19-somite stage. Various protective membranes and a substantial part of the outermost tissue layer, the ectoderm, have been dissected from this embryo so that underlying structures can be examined in the scanning electron microscope. The developing neural tube (spinal cord) runs from left to right and is bordered by a series of blocks of mesoderm cells called somites. The population of cells resting directly on top of the neural tube is the neural crest; as reflected here by their distribution, neural crest cells leave their site of origin and migrate away from the neural tube to take up residence at a variety of sites throughout the embryo. To the right, for instance, head ectoderm has been removed to reveal underlying head mesenchyme cells, most or all of which are of neural crest origin. A portion of the heart is seen below and to the left of the head. Above it is the cavity leading into the developing ear, and above that the dorsal surface of the hind brain has been broken away to permit viewing (from other angles) the inner surface of the developing central nervous system. (Courtesy of K.W. Tosney.)

TISSUE INTERACTIONS AND DEVELOPMENT

Norman K. Wessells
Stanford University

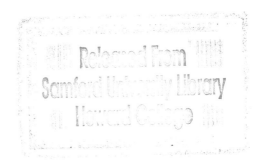

W.A. Benjamin, Inc.
Menlo Park, California • Reading, Massachusetts
London • Amsterdam • Don Mills, Ontario • Sydney

Book and cover design by Laurence Hyman
Drawings by Fran Milner

W.A. Benjamin, Inc.
2727 Sand Hill Road
Menlo Park, California 94025

Preface

This book is about the *process* of development. The emphasis is on *causes*, not on descriptions of embryos and the stages they pass through. The subject matter is chosen to supplement standard texts in embryology and developmental biology. It intentionally concentrates on the interface between traditional embryology and molecular biology. By learning how cells and tissues behave and interact in embryos, the reader will gain wider perspective for study in the first of those areas, and will better appreciate where and how the powerful tools of modern molecular biology can be used most advantageously in cracking the development nut. As a supplementary text in development, the book is written at a level appropriate for students who have completed a one-year course in introductory biology.

A comprehensive survey of the development of all organ systems is not our concern here. Instead, this book is a selective review and interpretation which treats most major principles of tissue interactions. For the sake of brevity, each concept, each line of experimentation, is not traced to its origins; hence, original credits are not given. Apologies are extended to the many scientists whose names are not mentioned; the omission is not from lack of regard but for ease in communication and conciseness.

Nevertheless, the most important part of this book for some readers will be the reference section at the end of each chapter. The cited

v

vi   books, reviews, and primary scientific papers were chosen to permit efficient access to the literature. Thus, original papers on given subjects are omitted if more recent ones of high quality and a wider range of references are available. Because so many of the best review articles are published in expensive books or medical journals unlikely to be available in some biology libraries, references to widely circulating journals are included, though they may not be the single, best treatment of that subject. The reference sections are fully annotated so that the reader can judge potential usefulness from those remarks, not from the original titles, which may be less informative.

One purpose of the book is to give the student a familiarity with the state of the art in this area of developmental biology. Limits in techniques and in theory are discussed, and areas of promise are pointed out. It is hoped that the text will help students develop the capacity to determine whether a scientific question is worth asking. In the days of seemingly unlimited funding and personnel, the shotgun approach to biology yielded substantial knowledge. Today, more taste must be displayed. Because a question can be asked at the molecular level does not mean it is worth asking, any more than just another histological or electron microscopic study of a developing tissue or cell is justifiable. Similarly, the fact that an enzyme activity or developmental process can be measured and assigned a number does not mean that it is more valid or valuable than less quantifiable phenomena characteristic of higher orders of biological organization. Our knowledge of tissue interactions in embryos is still so primitive that investigations at all levels are necessary if we are to fully explain these processes and their consequences in mechanistic terms.

My gratitude goes to Merton Bernfield, James Weston, John Philip Trinkaus, Robert Nuttall, and others for reading materials used in the book. The reviewers, Fred Wilt and Joseph Dickinson, contributed important organizational and factual improvements. Particular thanks are expressed to Margaret Moore, the production editor, Fran Milner, the artist, and Sharon Willi, the indexer, whose efforts have improved the book greatly.

The manuscript was prepared in part at the Stangeways Research Laboratory, thanks to the hospitality of Michael Abercrombie and support from the John Simon Guggenheim Foundation. Research mentioned in the text that comes from the author's laboratory has been generously supported by the National Institutes of Health and by the National Science Foundation.

The book is dedicated to two compatriots: the late Edgar Zwilling who, in 1970, urged me to prepare this brief treatment of tissue inter-actions. Zwilling was a warm friend, an exacting teacher, and a dis-passionate and self-critical scholar who stood as a model for all those fascinated by embryos. And, it is dedicated to J. P. Trinkaus for teach-ing me science and leading me to the good life of the professor.

Norman K. Wessells

Contents

A classic site of tissue interactions in a chick embryo: the neural tube above, the notochord below, and a somite to the right. The diaphanous cobweb of collagen and of protein-sugar complexes (see Chapter 15) extends from one tissue to another at this time when critical interactions are occurring between these tissues.

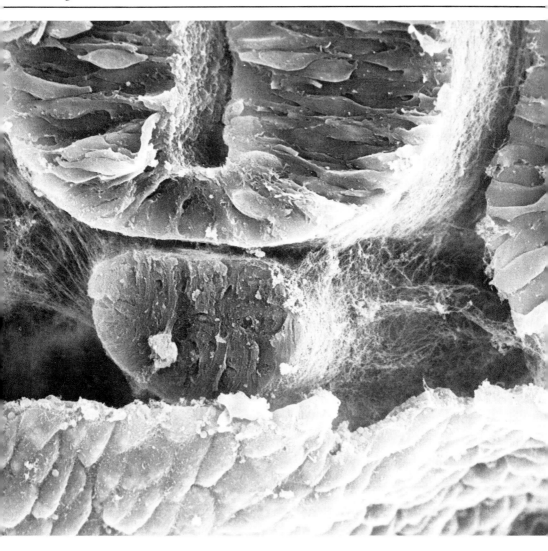

Orientation

Developmental biology is concerned with the multitude of processes that take place during the origin, maturation, and even senescence of the individual organism. Eggs, embryos, adolescents, and tissues that mature throughout life are among the objects of study. The fundamental goal of developmental biology is to explain how various types of information (information from genes, cytoplasmic substances, relative positions, physiological and chemical gradients, the environment) are integrated to allow a spherical egg a fraction of a millimeter in diameter to become a starfish, a butterfly, or a human being.

Tissue interactions, events in which dissimilar cell populations act on one another to alter cell behavior in developmentally significant ways, are central to that process. Virtually every organ in the adult body of many species arises as a result of tissue interactions. Eye, lung, kidney, leg, each organ would not be present if interactions did not

4   occur. Though present in cells, the genetic information normally employed in organ morphogenesis, cellular differentiation, and production of organ-specific proteins cannot be used if tissue interactions are absent or abnormal.

Tissue interactions are, therefore, one of the means by which genetic activity is regulated. As such, tissue interactions are both a source and type of "information" that is essential for development. But the means by which tissue interactions exert this regulation may be highly complex and indirect. Choice between pathways of cellular differentiation, mitotic activity, and morphogenesis are some of the diverse developmental processes that are affected by tissue interactions. One must anticipate that a variety of mechanisms operate during tissue interactions to allow control of such distinctive activities.

The purpose of this book is to explore typical tissue interactions, to consider possible mechanisms of interaction, and to assess the consequences of interactions for the cell, organ, and organism. After presenting a conceptual framework, we shall examine the development of two organ systems, the skin and the limbs, to establish some basic properties of interactions. We shall then proceed to a more detailed treatment of interactions in relation to morphogenesis and cellular differentiation. Next, we shall turn to types of interactions that continue in adult life, involving systemic hormonal action, the nervous system, and inhibitory controls of cell behavior. Finally, we shall consider some of the ways in which cells can affect one another and so participate in the tissue interactions that are the key to animal development.

The reader will find that the chapters focus on interactions. If this book were, instead, intended to provide a general treatment of embryology or a discussion of cell biology, then cell membranes might be treated fully in one place, chromatin in another, and so on. Examples from different organ systems might be mentioned in each such place. Here, the focus is different. Chromatin, cell surfaces, extracellular molecules are brought in wherever they are appropriate to illuminate our basic concern: interactions and their mechanisms. Furthermore, each chapter is built in part on preceding ones; consequently, a sequential reading of the book will be the most useful way to grasp the whole subject.

Perhaps the most important thing to be said as a conclusion to this orientation is a warning. The reader might be anticipating that we will attempt to define "the inducer," "the" agent that one tissue supplies to another to control its destiny in development. But every developing cell has a unique position, and is bathed in a spectrum of nutrients and ions that are dissolved in a solution whose osmotic pressure, pH, gas content, etc., all vary. Depending on one's bent of mind and means of analysis, any one part of that complex might appear to be limiting and so be "the" regulator under those conditions. Is there any hope of understanding systems with so many variables? For our purposes, we will consider many aspects of the environment of developing cells to be "background" that is essential, but not likely to be the source of regulation, as far as we can guess from current knowledge.

In addition, we will turn our attention to phenomena that are linked to a very important dimension of development: time. Development in embryos is most often a one-way street; that is, if one event aborts, then subsequent events are almost certain to be abnormal also. With this in mind, we can put physiological modulations—reversible changes in synthesis, responses to environmental variables, etc.—on the back burner with pH, osmotic pressure, etc. The remaining processes, which operate at unique times in cell and tissue development, will receive our attention as being the most likely candidates for control factors. Thus, even though this approach lets us stumble through the maze of variables and occasionally yields answers that look promising, we must never forget that our method is arbitrary, or forget the true complexity of the embryo as a living system that changes with time.

REFERENCES

N.J. Berrill and G. Carp. 1976. *Development.* McGraw-Hill. A detailed treatment of classical embryology and modern developmental biology; it includes useful sections on methodology. It also amplifies many of the topics treated in this book.

E. Deuchar. 1975. *Cellular Interactions in Animal Development.* Wiley. A comprehensive treatment of interactions in a wide variety of organisms, including many not discussed in this book. This is the most critical of the many recent European books of this type.

J.P. Trinkaus. 1969. *Cells into Organs.* Prentice-Hall. A scholarly, easy-to-read evaluation of the cellular basis of morphogenesis. Cell locomotion, guidance, and social behaviors are analyzed and provide an excellent complement to the discussions in this book.

J. Lash and J.R. Whittaker. 1974. *Concepts of Development.* Sinauer. A series of chapters of varying quality on topics that range from gamete formation to regeneration; good reference lists.

P.F.A. Maderson. 1975. *Amer. Zool., 15,* 315. This is an imaginative essay that links tissue interactions, modern cell and molecular biology, and evolutionary aspects of development.

Chapter One:

A ciliated cell on the outer surface of a salamander embryo. Ciliated cells like this one are scattered over the body surface of the embryo. Though such cells are spatially separated from each other, there is coordination between the cells so that the cilia beat in precisely specified directions to generate flow of water past the embryo. (See also, frontispiece to Chapter 16.) (Courtesy of P. Karfunkel.)

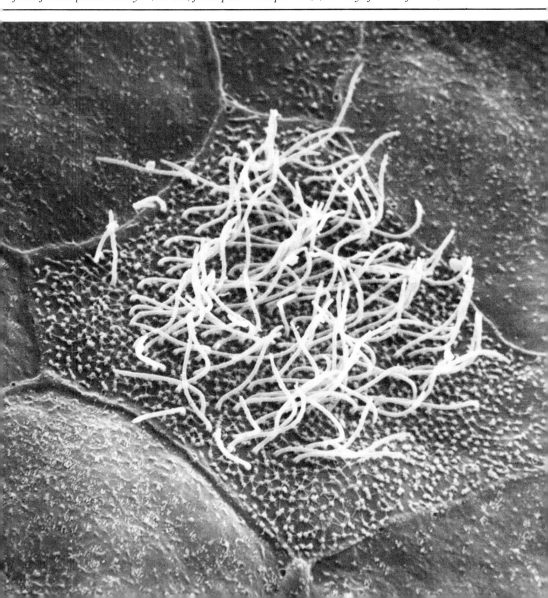

Conceptual Framework
of Development

*Some basic rules and definitions about cell division, genes,
and regulation.*

Most kinds of animals reproduce sexually. Despite the obvious differences between starfish, snails, butterflies, and human beings in the details of sexual reproduction and of their life cycles, a basic pattern is common to the lives of all such creatures.

The common thread of animal development involves the stages in the life cycle (see Figure 1.1). Eggs and sperm, arising by complex maturational processes termed *gametogenesis*, fuse at fertilization to begin development of the new individual.

Cleavage, a special type of cell division, increases the number of cells.

Gastrulation, a rearrangement of many of the new cells, brings cell populations into new spatial relationships with each other.

Organogenesis then occurs, and the major organs and tissues of the body form and begin to function.

8

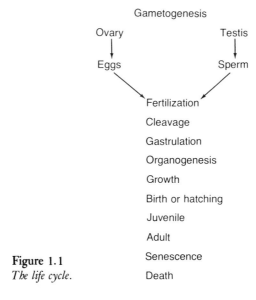

Figure 1.1
The life cycle.

The embryo (at least, most embryos) then grows to a size that will permit it to survive outside its protective layers (as, the eggshell) or environment (the mammal's uterus).

Birth or hatching then takes place, and the juvenile enters into one of a number of diverse forms. Larvae, pupae, tadpoles, infants, and teenagers share the common features of growth and maturation of body morphology and of physiological function. Evolution has, in a sense, played most freely with this phase of the life cycle.

Next, the adult state is reached. Maintenance of tissues and cell types is the key process. The gonads mature, carry out gametogenesis, and the individual organism reproduces.

Finally, at the conclusion of the reproductive phase, or after the rearing of offspring is finished (for species which perform that task), senescence and death complete the life cycle.

If we now look back over this list of phases in the life cycle, the most unexpected phase is gastrulation. If you were going to design a developing embryo, why include that gross and complicated rearrangement of cell populations? Why not instead put the cell populations that arise from cleavage in appropriate relative positions so that they can interact directly and carry out organogenesis? One reason for the universality of gastrulation movements may be that it is an

evolutionary remnant. The change from the two-layered blastula to   9
the three-layered gastrula may be a simple recapitulation of an equivalent evolutionary change early in the history of multicellular animal
organisms. In addition, and of more consequence for the main subject
matter of this book, cell movements at gastrulation and at later times
of embryonic development set up the conditions for tissue interactions.

We will turn to those interactions after we examine some background information about cleavage and other aspects of developing
cells.

The Egg, Cleavage, and Distribution of Substances

The mature animal ovum (egg) is a complex, differentiated cell. It
matures during the course of weeks or months, and comes to contain
a variety of substances (such as yolk, the food substance in eggs) that
are manufactured by the egg itself or derived from cells in the maternal

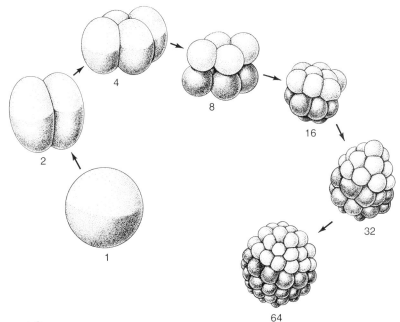

Figure 1.2
Cleavage in an organism such as a sea urchin. The division of preexisting mass is the
dominant theme of this phase of development. (After C.W. Bodemer, Modern
Embryology.)

10 body. Frequently these substances are arranged in special patterns within the egg cytoplasm, a phenomenon that holds great significance for development. Many eggs, when mature, do not actively carry out respiration, nucleic-acid synthesis, and other kinds of syntheses. Then most eggs undertake a final maturational process when ovulated or fertilized. This process may involve the completion of meiosis, the chromosome reduction divisions for the female genome, or the activation of various types of synthesis.

When the plasma membranes of egg and sperm fuse at the time of fertilization, many chemical reactions are initiated or accelerated. There may also be a highly ordered rearrangement of substances in the egg cytoplasm that we shall discuss in more detail below. In addition, events are started that culminate in the first mitotic cell divisions.

This series of divisions, called cleavage, is unusual in several ways. First, during early divisions just after fertilization, there is no so-called G_1 phase. In adult cells the cell cycle can be divided into four portions (see Figure 1.3): G_1 (gap 1); S (DNA synthesis); G_2 (gap 2); and M (mitosis *per se*). The G_1 phase is a major part of *interphase*, and may be a time of intense synthesis of RNA and of protein, and of net increase in cell mass before a new round of DNA synthesis *(S)* begins.

Now, let us return our attention to cleavage. Here, we find that G_1 tends to be absent or abbreviated. In other words, very little time is spent in "growing" during cleavage stages. Quite to the contrary, the hallmark of cleavage is division of the preexisting egg substances into separate cells or blastomeres. Since the total mass of most embryos does not increase during cleavage, but the number of cells goes from one to the thousands, it should be obvious that the egg is being broken into smaller and smaller packets, known as *blastomeres*.

One aspect of this process is that there is little synthesis of new ribosomes or of new types of proteins during cleavage; the ones that are already present in the egg at the time of fertilization suffice to carry the embryo through its early development. For most animal embryos, it is only at the time of *gastrulation*, the process of rearrangement of cell populations, that new types of proteins and various types of RNA must be made, if development is to continue. As would be expected, a G_1 period can then be measured during the division cycles. (In mammalian embryos, such syntheses may occur earlier, probably because of the relative absence in yolk and of the need to speed establishment of relations with the mother's uterus.)

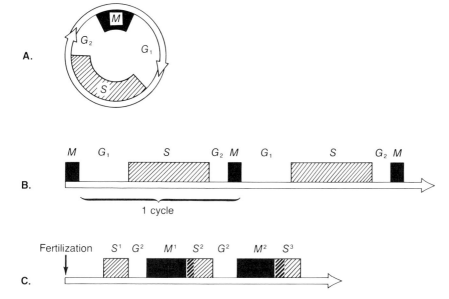

Figure 1.3

The cell cycle. A. The phases of the cell cycle in a typical late embryonic or adult cell. G_1, in particular, may vary greatly in length in different cell types. B. A linear representation of typical somatic cell cycles in Chinese hamster cells; note the relatively lengthy G_1 and S periods. C. The first cleavage cycles after fertilization in the sea urchin, Strongylocentrotus purpuratus. *Note that* S *begins before* M *is complete, thus completely eliminating a detectable G_1 period.* S *is also surprisingly short in comparison with* M *(M includes the classical phases of mitosis: prophase, metaphase, anaphase, telophase). (Data from R. Hinegartner et al. 1964.* Exp. Cell Res., *36, 53.)*

A second aspect of cleavage that deserves emphasis concerns the distribution of portions of the egg cytoplasm. Imagine that a fertilized egg has four types of special cytoplasmic materials, each occupying a patch on the cell surface (see Figure 1.4). The first two cleavages could segregate those patches among the four blastomeres. One result would be that only the progeny of each blastomere would contain the respective material. Now imagine that one of those materials allows certain genes to be used so that organ *A* forms; another causes other genes to be used so that tissue *B* develops; and so on. It follows that the initial distribution of the pigments relative to the planes of the cleavage divisions is crucially important, if the right organs are to form and are to be arranged in the embryo in a normal pattern.

We see therefore that the inclusion of substances, their initial distribution, their final placements after the rearrangements that follow fertilization (referred to above), and the manner in which substances

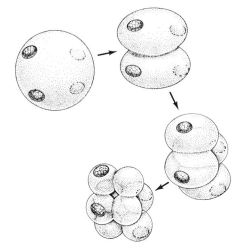

Figure 1.4
An imaginary egg with four patches of cytoplasmic information in its cortex. The planes of cleavage segregate the patches into individual blastomeres, each of which might be thought of as the starting point for a separate cell lineage.

are sequestered into blastomeres during cleavage can all have great consequence for development.

The inclusion of developmentally significant substances in eggs and their parceling out during cleavage provides a kind of "information" to the embryo. This sort of information is just as important to development as the genetic material of chromosomes is. Without it, correct regulation of the genome does not occur. In later chapters we will discuss other types of nongenetic information that are also crucial for development.

All Genes in All Cells

A second major concept of development concerns the genetic information available to developing cells. A series of observations begun during the 1880s on insects and worms demonstrated cases of chromosome "diminution" during development. Actual pieces of chromosomes were lost during the maturation of somatic cells in the organisms studied. These and other observations raised the possibility that the basis for cell differentiation may be such a process. Thus, in cells that will develop into mesoderm, the genes for ectodermal and endodermal tissues might be lost or destroyed.

In fact, this possibility seems unlikely, since later work has proved chromosome diminution to be the exception, not the rule. John Gurdon and his associates, following the pioneering studies of Briggs and

PREPARATION OF
DONOR CELLS

INJECTION OF DONOR NUCLEUS
INTO RECIPIENT EGG

PREPARATION OF
RECIPIENT EGG

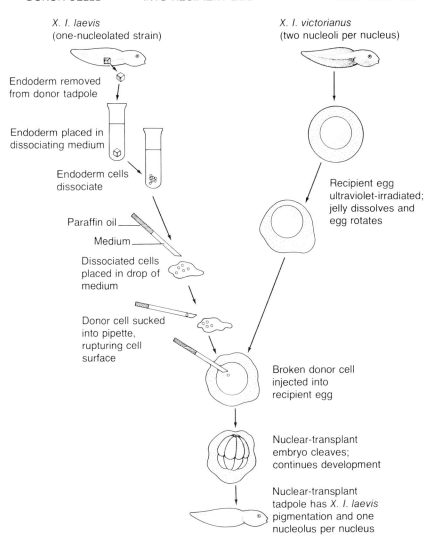

X. l. laevis
(one-nucleolated strain)

X. l. victorianus
(two nucleoli per nucleus)

Endoderm removed
from donor tadpole

Endoderm placed in
dissociating medium

Endoderm cells
dissociate

Paraffin oil

Medium

Dissociated cells
placed in drop of
medium

Donor cell sucked
into pipette,
rupturing cell
surface

Recipient egg
ultraviolet-irradiated;
jelly dissolves and
egg rotates

Broken donor cell
injected into
recipient egg

Nuclear-transplant
embryo cleaves;
continues development

Nuclear-transplant
tadpole has *X. l. laevis*
pigmentation and one
nucleolus per nucleus

Figure 1.5
The scheme for nuclear transplantation. Different subspecies of Xenopus laevis *are used, in order to prove that all differentiated cells in the final tadpole contain nuclei derived from the donor cell type. Instead of using endoderm from tadpoles (upper left) for donor nuclei, cultured adult cells or cancer cells might be employed. (Redrawn from J.B. Gurdon, 1966.* Endeavor, *25, 95.)*

14   King, have shown that a complete genome must be present in at least some fully differentiated cells. The nucleus and a small amount of cytoplasm from adult skin cells or larval intestinal cells can be injected into a fertilized egg of the clawed toad *Xenopus* that has previously had its nucleus destroyed by irradiation (see Figure 1.5). The injected nucleus then swells, comes to resemble a normal egg nucleus in morphology, commences the nucleic-acid syntheses that are appropriate for an egg nucleus, and then proceeds to participate in the cleavage divisions. Ultimately, a fully functional tadpole results, one that can be raised to adulthood and used for breeding.

Since the implanted nucleus is the sole source of nuclear genetic material in such an organism, it follows that the adult skin cell must have had a complete complement of genes, including ones which code for the proteins of muscle, nerve, blood, and other types of cells. Furthermore, although those genes would never normally be used in a differentiated skin cell, they are in a condition in which they can be reactivated as a result of passage through the egg cytoplasm, the cleavage process and so forth.

Let us assume, therefore, that a complete complement of genetic material is present in all adult cells (we will neglect special cases, such as red blood cells of mammals, that lack nuclei). If this is the case, then there must be a mechanism that selects the genes to be expressed in the functioning of the cell, since each differentiated type of cell is believed to contain a unique spectrum of proteins (hemoglobin is only present in red blood cells, insulin in pancreatic B-cells, etc.). The means whereby a given set of genes is selected for use in each type of cell is a primary problem of developmental biology. We shall discuss the degree to which tissue interactions may be involved in this selection process.

Proteins and Regulation

We shall be concerned with the criteria one uses to define the state of differentiation of developing cells. One set of criteria focuses on the types of proteins present in cells. According to Rutter's terminology, some proteins are found in all cells (the *primary proteins*), examples being enzymes needed for oxidative metabolism or for nucleic-acid synthesis. Other proteins occur in only a few types of cells *(secondary proteins)*, such as arginase which has been measured in liver, kidney, blood, and brain, but not in other tissues. Finally, *tertiary proteins* (which are frequently called *specific proteins*) are restricted to only one type of cell, hemoglobin in red blood cells, ovalbumin in oviduct cells, and so on (see Figure 1.6).

This difference in distribution may be related to how genes that code for those proteins function. Recently developed techniques enable us to assay the mRNA coding for a few of the tertiary proteins. This permits an independent check on how active genes are during protein synthesis, and also provides an independent criterion for the "state" of cellular differentiation.

Now, suppose you were asked to define the state of differentiation of a cell type. Would you choose a primary protein or its mRNA to measure? Probably not, since such proteins are distributed among many cell types and may be present at all stages of development. Instead, tertiary proteins or their mRNAs provide a much more sensitive assay, since they would be unique to the cell type under investigation. It is surprising how often this simple distinction has been ignored, and substantial time and research money wasted as a result.

Interpretation of measurement of tertiary proteins or their mRNAs must take into account a few subtleties that arise from the mechanisms which underlie protein synthesis. It is necessary, for instance, to distinguish between the levels of control of protein synthesis. Control can operate at the level of the gene. The gene is turned on, or not; but whenever it is "on," the mRNAs manufactured are used automatically to make protein. This form of regulation is termed "transcriptional" control of protein synthesis. Suppose, however, that an mRNA is made but is not used. It resides in the nucleus or in the cytoplasm in an inactive form. Then a separate signal (from that which caused the

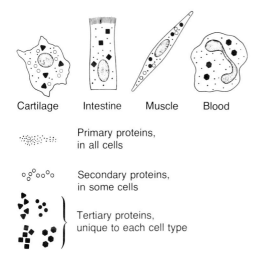

Cartilage   Intestine   Muscle   Blood

Primary proteins, in all cells

Secondary proteins, in some cells

Tertiary proteins, unique to each cell type

Figure 1.6
Rutter's scheme for protein distributions. A tertiary protein, because it is found in a single type of cell, provides the best criterion for the state of differentiation of a cell or cell lineage.

16

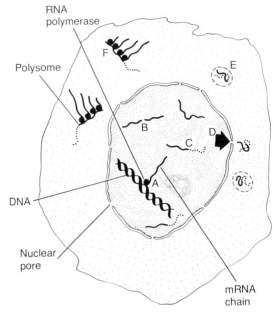

RNA polymerase

Polysome

DNA

Nuclear pore

mRNA chain

Figure 1.7
Possible sites of regulation of protein synthesis. (A) transcription—at the level of mRNA production from the DNA template. (B) nuclear RNA processing; cleavage to smaller pieces. (C) nuclear RNA processing; addition of bases. (D) nuclear RNA processing; control of egress from the nucleus. (E) translation—mRNA in cytoplasm but inactive. (F) active polysomes, the sites of polypeptide chain elongation using mRNA as a template. In fact, steps B through E are all included in the "translational" level of control.

mRNA to be made on the DNA template) starts actual protein synthesis using that mRNA. Here we say "translational" control has operated. The possibility that any given differentiating cell might utilize the translational control mechanism, suggests that assays of mRNA levels alone may not be the best criteria for the state of differentiation of cells. This is because differentiation has to do with *function*, with the actual enzymes or structural proteins that permit cells to function in their unique ways. Consequently, measurement of proteins, not of the messengers that may or may not be translated into proteins, gives a more accurate assessment of functional phenotypes of cells.

Restriction and Expression

Following cleavage, embryonic cells become segregated into subpopulations that we recognize as incipient organs and tissues: the medullary plate is the precursor of the vertebrate central nervous system; the limb bud the forerunner of the arm or leg; the ureteric bud the precursor of kidney; and so on.

What do we mean by "segregation" into subpopulations? Only certain cells of the covering layer of the embryo become the prospective nervous system. The rest are destined to form skin, hair, glands, or other organs. Clearly, something must enable such choices to be made. Tissue interactions between cell populations are a major source of information for this segregation process.

Note that we have been talking about segregation into "incipient" organs. Many distinctive processes occur during the subsequent development of the precursor populations into the final organs. These processes may also largely depend on tissue interactions.

Therefore, to avoid confusion in our analysis of development, we must carefully distinguish between events that set up populations of cells which will be committed to various types of development, and the events of morphogenesis and cellular differentiation, in which the cells actually build organs and tissues. For this reason, we will assume that each subpopulation of cells in an embryo passes through two major maturational phases, which we shall call the *restrictive* and *expressive* phases of development. The restrictive phase concerns commitment and segregation, whereas the expressive phase concerns expression of the genome and other forms of information as organs and tissues are constructed.

The two terms will be used in what the experimental biologist calls the "operational" sense. Restriction, determination, and the various kinds of expression are recognized by means of specific experimental operations and measurements we employ (these will be explained in following chapters). We must keep those operations in mind when we consider restriction and expression, for in a very real sense the operations define the terms. Terms such as these are only valuable to us if they simplify the complexity of cell behavior in development, and can suggest what sorts of experiments and measurements will be most useful in ultimately explaining that behavior in chemical and physical terms.

18   The operational definition of tissue interactions themselves is also worth emphasizing here. What we will consider in the following chapters are interactions between populations of cells. The state of our experimental art does not yet permit us to answer our questions for individual cells; that is, we cannot ask whether *a* mesoderm cell acts on *an* epithelial cell to cause its maturation. Instead, we demonstrate that a mesoderm population (with its complex intercellular environment) interacts with an epithelial population (with its special properties), and the result is morphogenesis. In fact, it is not far-fetched to suppose that some of the responses to tissue interactions cannot be seen or measured for single cells, since they constitute a multicellular activity. These strictures limit the precision of our statements and our conclusions about tissue interactions.

CONCEPTS

Cleavage divisions divide the egg into smaller and smaller packets, the blastomeres.

Substances in the zygote surface or in its cytoplasm can be sequestered into restricted numbers of blastomeres because of the planes of the cleavage divisions.

Some substances distributed in this way can affect or limit the type of differentiation carried out by cells.

The complete genetic complement is present in nuclei of most types of differentiated cells.

Proteins (and mRNAs) show varying types of distribution among cell types. Some are found only in a single kind of differentiated cell.

Regulation of gene activity may occur at the transcription or translation steps that lead to protein synthesis.

Restriction refers to limitation in the ways that a cell population can differentiate.

Expression refers to use of genetic and other information during morphogenesis and cell differentiation.

REFERENCES

Early development:
American Zoologist, vol. 15, no. 3 (Summer, 1975). The volume treats many aspects of early development in Echinoderms, and includes a recent summary by D. Epel of biochemical events that follow fertilization; p. 507.

Cleavage:
R. Rappaport. 1974. In J. Lash and J.R. Whittaker, eds., *Concepts of Development*. Sinauer. This paper is an excellent entry to the literature, and presents an imaginative analysis of the cleavage phenomenon.

Cell cycle:
J.M. Mitchison. 1971. *The Biology of the Cell Cycle*. Cambridge Univ. Press. Methodology, theory, and practical examples about cell cycles.

Nuclear transplantation:
J.B. Gurdon. 1974. *The Control of Gene Expression in Animal Development*. Harvard Univ. Press.
J.B. Gurdon. *Scientific American*, December 1968, p. 24.
M.A. DiBerardino and N. Hoffner. 1970. *Develop. Biol., 23,* 185.
These papers present both sides of the nuclear-transplantation debate, and summarize Gurdon's extensive experiments on mRNA injection into eggs.

Proteins and regulation:
W.J. Rutter, R.L. Pictet, and P.W. Morris. 1973. *Ann. Rev. Biochem., 42,* 601. This is a general review of differentiation mechanisms that distinguish levels of regulation.
P.A. Marks and R.A. Rifkind. 1972. *Science, 175,* 955.
J. Ross, I. Ikawa, and P. Leder. 1972. *Proc. Natl. Acad. Sci., 69,* 3620.
These papers are on hemoglobin mRNA and genes during early red-blood-cell differentiation.

Chapter Two:

A fruit fly (Drosophila) *embryo near the completion of cleavage and just prior to gastrulation. The pole cells are seen at the upper end of the embryo. Ultimately, the pole cells will reside within the gonads of this fly and give rise to its germ cells (sperm or eggs). (Courtesy of F.R. Turner and A.P. Mahowald. From* Develop. Biol., 50 (1976), 95.

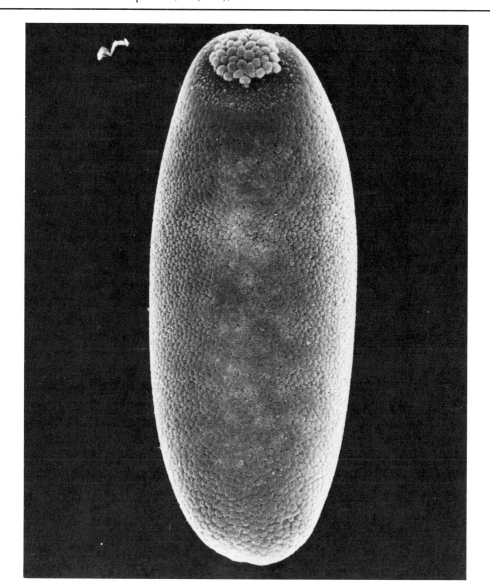

The Restrictive Phase

Pigment cells divide again and again and again, but remain pigment cells. Why?

Suppose we challenge certain cells at the anterior end of a human embryo at various times in their maturational history by using experimental techniques that pose the question: how many potential tissue types can cell group X form? Figure 2.1 summarizes results we might obtain. Early in the history of our cell population, the capacity to form a variety of mature cell types is present. Gradually the latitude decreases, until a final commitment is made—only lens cells can be formed. We call the loss in ability to form a variety of cell types *restriction;* and the final step to a single commitment, *determination*.

The timing of restriction is important for certain types of tissue interactions. The time at which restriction occurs is highly variable. In some cell lines, the restrictive phase may occur soon after fertilization, as specific substances present in the original fertilized egg are

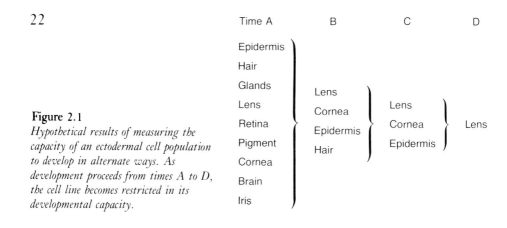

Time A	B	C	D
Epidermis			
Hair			
Glands	Lens		
Lens	Cornea	Lens	
Retina	Epidermis	Cornea	Lens
Pigment	Hair	Epidermis	
Cornea			
Brain			
Iris			

Figure 2.1
Hypothetical results of measuring the capacity of an ectodermal cell population to develop in alternate ways. As development proceeds from times A to D, the cell line becomes restricted in its developmental capacity.

Figure 2.2
Where did those ectoderm cells go? Down into the hole to form the lens. The cells remaining on the surface will cover over the hole and give rise to the cornea of this chick embryo's eye. These are the kinds of cells undergoing the restriction and determination processes described in the text. (Courtesy of K.T. Tosney.)

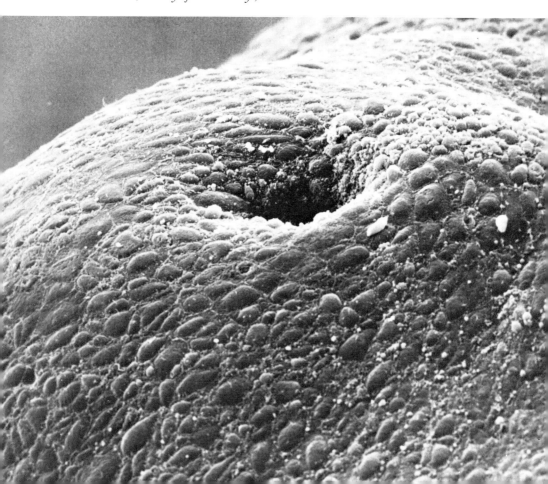

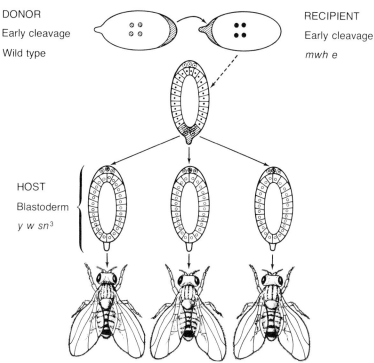

Figure 2.3

An experiment to establish the significance of the pole plasm. Pole plasm is removed from its normal posterior position in an egg of one strain of Drosophila. *The plasm is injected into the anterior end of a host egg of another strain. Subsequently, primordial germ cells appear in that anterior region; they are then transferred to still another genetic line of fruit fly. When those final host flies mature, they can be mated and the progeny checked for the presence of genes found in the original recipient* (mwh e, *in the diagram). In fact, such genes are present and expressed. This shows that the pole plasm caused cells located at the anterior end of the first recipients to become primordial germ cells which, in the final hosts* (y w sn³), *could give rise to sperm or eggs. (Redrawn from K. Illmensee and A. P. Mahowald,* Proc. Natl. Acad. Sci., 71 *(1974), 1016.)*

parceled out during cleavage. For example, in insect eggs, a substance called "pole plasm" is found in one area of the cytoplasm (see Figure 2.3). During the cleavage divisions, this pole plasm is incorporated into certain cells, which are thereafter committed to form "primordial germ cells" (the sources of sperm and eggs in mature gonads). The same kind of process goes on in some and perhaps all vertebrates. In amphibian embryos, for instance, particles rich in RNA and protein

24   are segregated during cleavage into certain ventrally situated blastomeres that later give rise to the primordial germ cells of ovaries or
testes. If the pole plasm or RNA-rich granules of amphibians are
irradiated with ultraviolet light at wavelengths that destroy RNA,
then sterile gonads result. No substitute primordial germ cells can
form—even though the rest of the testis or ovary is normal, correct
sex hormones are present, and so on—because the information required for primordial germ-cell development has been eliminated by
our treatment. Once gone, it cannot be regained, even though the com-

Figure 2.4
*The primordial germ cells (pole cells) of a fruit-fly embryo at an earlier stage than in the
frontispiece of this chapter. Note the two cells that appear to be dividing (M). Later in
development, at the time of gastrulation, these primordial germ cells will be clustered together
as in the frontispiece. (Courtesy of F.R. Turner and A.P. Mahowald. From* Develop.
Biol., *50 (1976), 95.)*

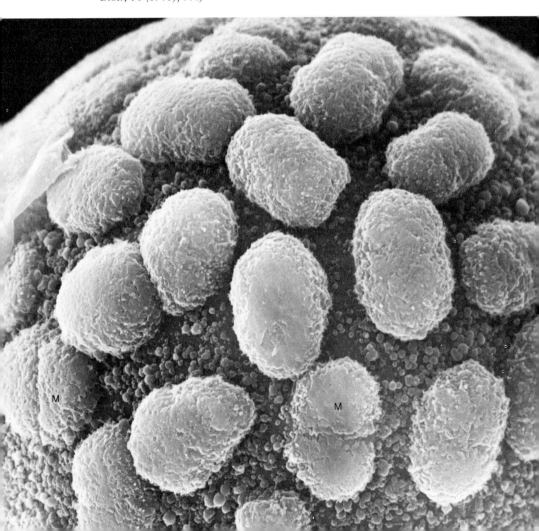

plete genetic information of the organism (including that used in primordial germ-cell maturation and function) is present in its gonadal cells.

The primordial germ cells are an example of early restriction, which depends on substances manufactured and stored during egg maturation. Once restricted, the primordial germ cells cannot form muscle, brain, kidney, nor even, we believe, the other (nongerm line) parts of gonads themselves (i.e., the hormone-secreting cells, nurse cells, etc.).

In contrast to the early restriction of some cell types, other cell lines may not undergo final determination until relatively late in development (as in Figure 2.1). Even in such cases, the initial stages of restriction may occur during cleavage, perhaps as the result of differential distribution of cytoplasmic factors of the egg. In amphibians, for instance, a "gray crescent" region appears on the surface of the egg soon after fertilization (see Figure 2.5). This crescent consists of the egg cell surface (the plasma membrane and attached substances, inside and outside that membrane). Next, cleavage commences and the crescent becomes segregated to certain of the blastomeres. Hours or days later (depending on the species and its rate of development), cells that derive materials from the crescent move to the interior of the gastrula and form a mesodermal cell population. If the gray crescent is removed from an early embryo any time before or soon after the time of first cleavage, such cell movements do not continue and no axial mesoderm appears (and the central nervous system, which forms in response to the axial mesoderm, also does not develop).

The gray crescent acts in a formal sense like the pole plasm. However, cells that derive materials from the crescent are not fully restricted. They retain much developmental plasticity, and can form muscle, cartilage, bone, or perhaps other cell types of the mesodermal family. The particular tissue and cell type that they do form will be governed by interactions that occur in the environment they ultimately will inhabit. The important point is that once they are restricted to forming mesodermal-type tissues and cells, they then cannot form ectodermal (brain, epidermis, etc.), or endodermal (stomach, liver, etc.) cell types. Thus, they are restricted but not yet fully determined.

Clearly, there are basic similarities between the precocious determination of the germ-cell line and the relatively tardy final determination of mesodermal lines. The main obvious difference between them is timing. Moreover, if we think carefully about the apparent stepwise

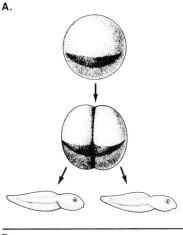

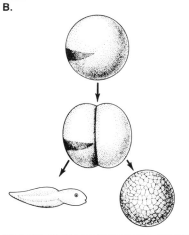

Figure 2.5
*Experiments on the gray crescent. **A**. Normally, the first cleavage furrow passes through the gray crescent. If the first two blastomeres are separated from each other after the first cleavage, both blastomeres can develop into a normal tadpole. **B**. If the first cleavage is caused to pass through the zygote so that all of the crescent is segregated into one blastomere, then, after separation of the blastomeres, only the one containing gray crescent materials can undergo gastrulation and develop normally. **C**. If a piece of gray crescent from one zygote is implanted in the surface of a host zygote, then two points of gastrulation infolding appear and a double tadpole would result. (After Curtis.)*

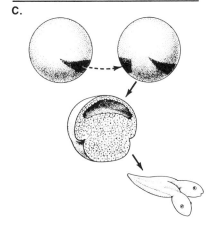

restriction of the mesodermal line (or ectodermal one, as in Figure 2.1), we can see that there is little reason to assign special attributes to the final determinative restriction. It appears special only because it occurs last. That does not mean, as some biologists have assumed, that its mechanism is necessarily unique (though it might be).

Before going on, let's be quite clear why we have discussed the relative timing of restriction. The pole plasm and gray crescent serve as models to illustrate that cytoplasmic factors can control cell behavior and contribute to the restriction process in early embryos. As we will learn, certain tissue interactions occurring much later in the history of the embryo also "cause" the latter steps in restriction and determination. Does this mean that such interactions operate via analogous cytoplasmic "factors"? That and other interesting questions come to mind immediately. What should now be kept foremost in mind is that not all the requisite information for restriction of all cell types is parceled out during early development. Tissue interactions that occur later are essential, if many kinds of restriction are to take place.

Significance of Restriction

What is the significance of the restriction-determination process? We can see if we take a determined cell line, say, of pigment or cartilage cells, and by using suitable culture nutrients, cause the cells to divide again and again. We find that the cells "breed true"; cell generation after cell generation, they remain pigment or cartilage in type. Thus *the restriction process results in stable, heritable limitations being placed on cells.*

These experimental results imply that every time the cells divide in our culture dishes, or determined cells divide in an organism, not only are new DNA, chromosomal proteins, and chromosomes manufactured, but also the new genome is kept in the same condition as the parental one. That is, certain genes apparently are permanently repressed (like ectodermal and endodermal ones in a mesodermal cell line); others are in a condition where they can be used if needed (like the cartilage or pigment genes in the example above); and still others are being used (like those concerned with general cell structures or processes). The means by which the heritability of the determined state can be maintained with little likelihood of error during one cell-division cycle after another is a major mystery of cell biology.

28   The great stability of the determined state is probably one of the most important attributes of mature cells, since it helps to increase the probability that, except for truly extraordinary circumstances (cancers, grave wounds, etc.), cells will remain true to type for the lifetime of the adult animal.

Cytoplasmic Substances, Chromatin, and Heritable Properties

Because of these considerations, it is of interest to discuss briefly some possible sources of heritable properties of nuclei. The so-called "o" mutants of axolotls (salamanders) provide evidence that a protein may be involved in one heritable property of nuclei. The o^+ substance is made in the nucleus of each developing egg of normal, wild-type female axolotls. The substance, probably a protein, is released into the egg cytoplasm when the nuclear envelope breaks down because of hormone action. Subsequently, after fertilization of the egg, the protein acts at about the midblastula stage on the nuclei housed within the individual blastomeres of the embryo. As a result, those nuclei will show altered patterns of RNA synthesis, protein synthesis in the cytoplasm will be accelerated, and cleavage and gastrulation will be able to proceed normally.

How do we know this? Homozygous mutant axolotl eggs (o^-/o^-) lack the o^+ substance and stop cleaving normally at the late blastula stage. Suppose we enucleate an activated o^-/o^- egg (recall the technique for such experiments from Chapter 1), and then inject into its cytoplasm an o^+/o^+ nucleus from an *early* blastula. Development aborts just as it would if the original mutant nucleus was present. Apparently an early blastula nucleus from a normal embryo has not yet been acted upon by the o^+ substance so that it can support development in mutant cytoplasm.

But suppose we inject an o^+/o^+ nucleus from a *late* blastula cell into the enucleated mutant egg cytoplasm (o^-/o^-). Development proceeds quite normally! It is concluded that the o^+ substance acts between mid and late blastula, and somehow alters nuclei so that they are unaffected by being engulfed in mutant cytoplasm.

Now, what of heritability? Suppose we place an o^+/o^+ *late* blastula nucleus in an o^-/o^- mutant enucleated egg and let cleavage proceed. After a number of cleavage divisions, we remove a nucleus from one of the blastomeres and inject it into a freshly enucleated o^-/o^- mutant egg. Cleavage proceeds. The process is repeated several times, so that the lineage of o^+/o^+ nuclei replicates many times in o^-/o^- cytoplasm. Then, if we let one of these embryos (o^+/o^+ nucleus, o^-/o^- cytoplasm) develop, all is normal—gastrulation proceeds.

What can be concluded? The original nucleus arose during cleavage in the presence of the o^+ substance in the original donor embryo. That nucleus had been *permanently* altered—heritably altered—so that it could participate in division again and again and again, but not lose the ability to support normal development.

These experiments do not tell us whether the o^+ substance works directly on the chromatin or whether it alters some other aspect of nuclei. Nor do they tell us that the great relative stability of the determined state is due to cytoplasmic proteins that act within or on nuclei (we are not even dealing with restriction or determination in these experiments). However, they do provide us with an important model about heritability. *The experiments show that heritability can be a property of nuclei themselves, and can stem from interactions of a cytoplasmic substance with nuclei.* Those are important conclusions for any general theory of development.

Next, let us consider the primary constituent of nuclei, the chromatin (DNA and associated proteins). Chromatin can be examined for its capacity to act as a template for the synthesis of mRNAs. Paul and others have shown that chromatin can be extracted from cells, purified, and used as an active template. Chromatin from a cell that is synthesizing globin (the protein of hemoglobin) allows transcription of globin mRNA, whereas chromatin from brain or other tissue types does not. Thus there is a restriction on transcription which operates at the level of chromatin and is *tissue-specific* or *organ-specific* in character. Some experiments suggest that the nonhistone proteins of chromatin account for this specificity in template activity (see Figure 2.6).

These are most important observations, since they imply that the stability of the determined state could, in part, be due to the structural organization of chromosomes. Whatever process accounts for the chromatin specificity, it is not irreversible. Recall the nuclear transplantation experiments that demonstrated unequivocally that some nuclei from differentiated cells can support all normal development. The chromatin and nonhistone proteins cannot be associated in an irreversible way or this result could not be obtained. Unfortunately,

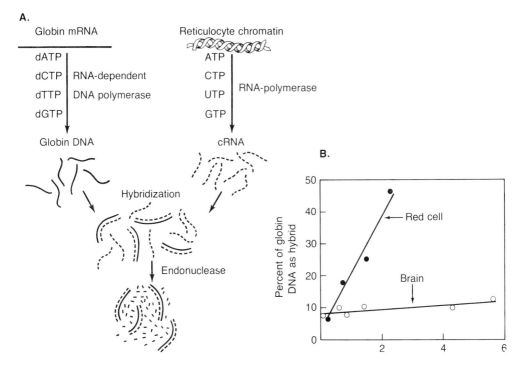

A.

Globin mRNA

dATP	
dCTP	RNA-dependent
dTTP	DNA polymerase
dGTP	

Globin DNA

Reticulocyte chromatin

ATP	
CTP	
UTP	RNA-polymerase
GTP	

cRNA

Hybridization

Endonuclease

B.

Percent of globin DNA as hybrid

Red cell

Brain

Figure 2.6

*A diagrammatic scheme of the method used in measuring the amount of hemoglobin mRNA manufactured using chromatin as a template. **A.** Chromatin from developing red blood or brain cells is mixed with RNA polymerase, precursors of RNA, and other cofactors. RNAs that are complementary (cRNA) to the chromatin are produced. Globin DNA is prepared by using highly purified globin mRNA as a template. When an RNA-dependent DNA polymerase, precursors of DNA, and cofactors are all present, globin DNA is made. Then the single-stranded globin DNA and the mixture of cRNA are allowed to "hybridize." An enzyme (called an "endonuclease") is then added that degrades single-stranded DNA; i.e., any of the globin DNA that is not in hybrid form with complementary globin mRNA. If, for instance, no globin mRNA was present in the cRNA mixture made on the chromatin, then all the DNA would be single-stranded and would be degraded. Thus, the amount of DNA that is not degraded because it is in hybrid form with mRNA is an indirect measure of that mRNA. **B.** Typical results of performing the assay using cRNA prepared from brain chromatin (open circles) and red-cell chromatin (dots). It is concluded that chromatin from the two cell types differs in its ability to make RNA that is complementary to globin DNA. (From the papers by Axel and by Paul.)*

these considerations also raise the question of what keeps DNA and nonhistone protein together in a specific pattern in a differentiated cell. Is it an intrinsic stability which can only be canceled by extreme measures, such as exposure to the very special environment within a fertilized egg? Or is it due to nuclear or cytoplasmic factors, which in a sense would act just oppositely to the cytoplasm of an egg? For these reasons we cannot assume that the ability of DNA, histones, and non-histone proteins to remain associated in a specific manner during isolation and purification procedures in test tubes necessarily reveals the source of the heritable stability of the determined state. We shall return to these questions in Chapter 13, after discussing the cell surface in development.

CONCEPTS

Restriction refers to the loss in ability of cells to differentiate in alternative ways.

Determination is the final step in restriction, and involves commitment of a cell line to a specific pathway of differentiation.

Restriction apparently can occur either as a series of discrete steps or as a single step.

The determined state is a heritable characteristic of cells.

Isolated chromatin displays tissue-specific transcriptional properties.

Cytoplasmic substances can cause heritable alterations in nuclei of early embryos.

32 REFERENCES

Germ plasm:

L.D. Smith and M.A. Williams. 1975. In, C.L. Markert and J. Papaconstantinou, eds., *Biology of Reproduction.* Academic Press. A critique and review of germ plasm and primordial germ-cell determination in insects and vertebrates.

Gray crescent:

A.S.G. Curtis. 1962. *J. Embryol. Exptl. Morphol.,* 10, 410. This is one of the most remarkable transplantation experiments ever attempted. It summarizes effects of gray crescent activity.

Cytoplasmic determinants:

J.R. Whittaker. 1973. *Proc. Natl. Acad. Sci.,* 70, 2096. This is the clearest case of cytoplasmic factors acting to control synthesis of specific enzymes during differentiation.

Distribution of cytoplasmic factors during cleavage:

G. Freeman. 1976. *Develop. Biol.,* 49, 143. An excellent recent avenue to the literature on this important topic.

Heritability of determination:

R.D. Cahn and M.B. Cahn. 1966. *Proc. Natl. Acad. Sci.,* 55, 106.

H.G. Coon. 1966. *Proc. Natl. Acad. Sci.,* 55, 66.

These are the two pioneering and classic papers proving the heritability phenomenon.

Chromatin transcriptional stability:

Cold Spring Harbor Symposium. Quant. Biol., 38 (1973). This volume contains several papers on this and related subjects, including two on hemoglobin chromatin by R. Axel *et al.* (p. 773), and by J. Paul *et al.* (p. 885).

"o" substances in axolotls:

A.J. Brothers. 1976. *Nature,* 260, 112. This is a recent summary of work on the "o" substances and gives references to the Briggs' groups' pioneering experiments in this field.

Chapter Three:

An electron micrograph of the cytoplasm of a developing muscle cell from a chick embryonic heart. Filaments composed of actin, myosin, and other proteins of the contractile apparatus are becoming aligned in functional units called sarcomeres (a single sarcomere stretches from S1 to S2). None of these structures were present in this cell when it began to differentiate. Many more will ultimately accumulate, and they will be regularly aligned, so that the cell will be able to participate effectively in generating the heart beat.

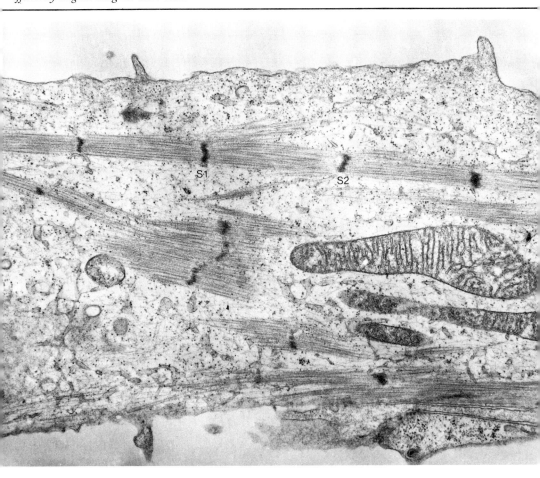

The Expressive Phase

Though the determined state is heritable, the cell's functional phenotype can vary, seemingly, at the whim of the tissue environment.

Let us now turn to the second, expressive phase of cell development. Imagine that we can observe a group of restricted cells in their normal environment in an embryo. At a specific time, the group might begin to form a hollow treelike structure, the lining of the lung. *Morphogenesis*, one type of expressive process, is going on. Then, as we follow our population, intracellular organelles become redistributed, secretory granules appear, an abundant rough endoplasmic reticulum accumulates, and the cells begin to function as they do in the adult. *Cell differentiation*, the construction of the final, functional cell phenotype, is occurring. This is the other major expressive process in development. And, although we may not be able to "see" it, an important capacity is acquired during this differentiation, namely, the ability

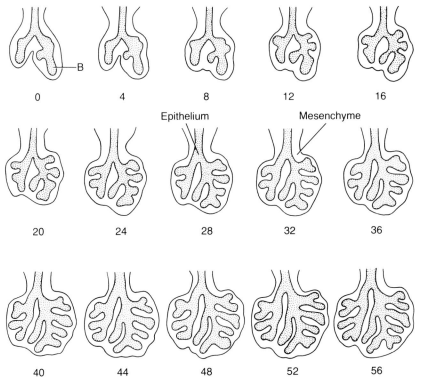

Figure 3.1

Tracings of photographs of a mouse lung system carrying out morphogenesis in culture. The epithelium (dark) branches extensively and grows in mass during this period of 56 hours. Note that the right and left halves of the lung have different shapes, and carry out slightly different branching programs. These differences are seen in every mouse lung system, suggesting a genetic basis. To appreciate the morphogenetic process itself, follow single bronchial buds (B) through the series of tracings. The surrounding mesoderm tissue (light) increases in mass, while, within it, smooth-muscle and blood cells develop in response to interaction with the epithelium. (After T. Alescio. La ricerca sci., 35, series 2, II-B, vol. 6 (1965), p. 237.)

to be regulated by the sorts of cues present in the adult organism. Thus the differentiated cell must be able to respond appropriately to the endocrine system, the nervous system, or whatever else serves as the external source of information that integrates the cell's function into the economy of the whole organism.

Now, we must consider one more aspect of "timing" in development, namely, when the expressive phase may start in relation to the restrictive phase. One extreme is illustrated by the mammalian pancreas, an organ we will consider in detail later. Expression, in the form of morphogenesis and of cell differentiation in the pancreas, follows

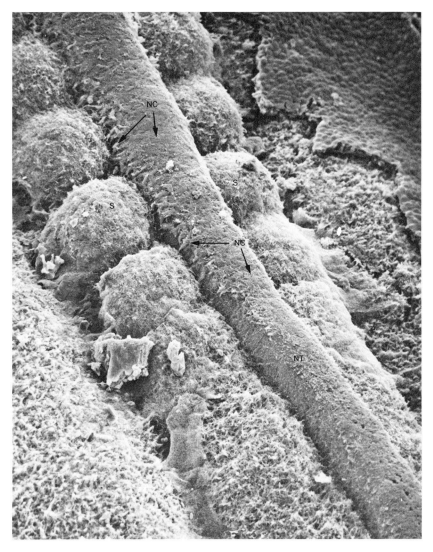

Figure 3.2

Morphogenesis of a different type than that described for various epithelia is seen in this view of a chick embryo. The ectoderm has been removed from the lower back to reveal the newly formed neural tube (spinal cord) extending from upper left to lower right (NT). The paired somites (S) are seen on each side of the neural tube. They arise by a process of morphogenesis in which the continuous mass of mesoderm on each side of the neural tube (lower right) breaks up into blocks of somite cells. Cells seen on top of the neural tube comprise the "neural crest" (NC). The varying degree to which those cells have spread toward adjacent somites, as one looks from anterior (top) to posterior along the neural tube, reflects the migratory activity of individual neural crest cells. Such cells will migrate to many sites in the embryo and give rise to a diversity of differentiated cell types, depending on the local environmental influences to which they are subjected. (Courtesy of K. T. Tosney.)

38   close on the heels of determination. Thus little time separates restriction from expression. But consider the mammary gland. The mammary epithelium apparently becomes fully restricted in the embryo, undergoes initial morphogenesis and differentiation, and enters a period of inactivity. Then, months or years later (depending on the species of mammal), changing levels of hormones during pregnancy finally elicit completion of epithelial morphogenesis and large-scale production of milk. This separation in time of restriction from terminal expression emphasizes the independence of the regulatory events involved in the one process from those involved in the other. One might guess, therefore, that the "information" which causes restriction and initial expression may be quite different from the hormonal spectrum that leads to final maturation and function. If this is so for the mammary gland, it also may apply in situations where restriction and expression are closely coupled in time, even though it is much more difficult for us to distinguish such independence because of the closeness in time of the different processes.

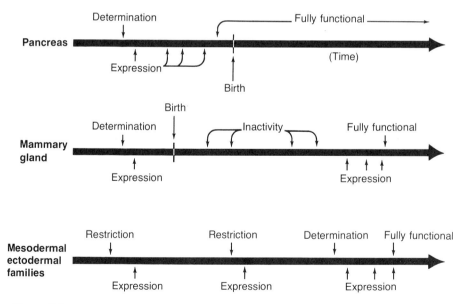

Figure 3.3
The relative times of restrictive and expressive events in various organ systems. In all cases, it is probable that at least some expression follows close after restriction and determination (e.g., initial morphogenesis, low-level synthesis of specific proteins).

We have chosen to discuss the pancreas and the mammary gland as examples of two extremes. In fact, when we survey the large number of organs that develop from cell populations that undergo a lengthy restriction phase it becomes clear that preliminary expressive events can take place even before final determination (see Figure 3.3). Thus the partially restricted mesodermal cells derived from the gray crescent area carry out a characteristic pattern of morphogenetic movements as they migrate to the interior of the embryo. Once inside, some of the cells may participate in forming a somite. Even so, depending on environmentally imposed factors, they can be caused to form cartilage, muscle, dermis, or perhaps other determined cell types. The point is that preliminary expressive processes can go on in cells at various stages of the restrictive phase. *The expression at any one time is appropriate to the restrictive state at that time.*

If this is so, we would not expect to be able to cause a gray crescent-derived mesoderm cell which is carrying out gastrulation movements to suddenly begin producing muscle-specific proteins. The cell, or its progeny, would have to pass through various maturational events, including determination as a myoblast, before such syntheses could be evoked. These cases emphasize that the initial steps of restriction can be followed by expressive activity, so that, step by step, the cell line matures. Restriction and expression are certainly not mutually exclusive processes.

The reason that we are going into these seemingly complex interrelationships is that many of the tissue interactions we will shortly describe involve control of both restrictive and expressive events. This will compound our task of explanation, since distinguishing separate control processes is very difficult.

Significance of Expression

An important difference distinguishes the stability of the final restricted (determined) condition from the stability of the final "expressed" state. By the latter term we mean the final differentiated "phenotype" of a cell. First, imagine that we alter the nutrients on cultured cartilage or pigment cells so that a rather crude mixture of undefined substances is present (see Figure 3.4). The cells cease making components of cartilage matrix and pigment granules. Their characteristic intracellular architecture is simplified as they undergo

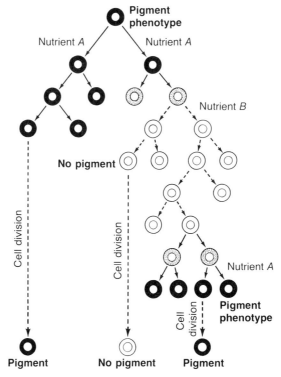

Figure 3.4

The stability of phenotype and of the determined state. As long as the cells are in nutrient mixture A, the phenotype is expressed. When they are placed in B, the phenotype is lost. On return to A, the cell line once again expresses its original character, reflecting the stability of the restricted condition. (See papers by Cahn and Cahn, and by Coon.)

repeated cell-division cycles. They have lost their differentiated "phenotype." Or we might wound the skin or break a bone in an embryo or an adult. Again, the differentiated phenotype of skin or bone cells is lost as the cells enter a state of mitosis and repair activity. The conclusion is that *the differentiated phenotype is remarkably labile.* It is a function of the normal tissue environment; if the environment is altered in certain specific ways, the phenotype will be lost. But in such cases, if the original nutrients are returned to the cultured cells (see Figure 3.4), or if the wound is repaired, the cells will return to their original differentiated phenotype, be it cartilage, pigment, skin, or

bone. This reflects the fact, discussed earlier, that the determined state
is relatively *stable*, and is not dependent on a constant environment
for its integrity.

CONCEPTS

Two of the major processes of expression are morphogenesis and
differentiation.

Morphogenesis involves changes in shape or position of cells,
tissues, or organs.

Differentiation involves maturation of cells into a functional condition (i.e., from the point of view of the organism).

The differentiated cellular phenotype is *not* a heritable characteristic of cells, but is maintained by properties of the local tissue
environment, control factors, etc.

REFERENCES

Differentiation:
C. Grobstein. 1964. *Science, 143,* 643. This seminal paper poses many of the
    important questions concerning cellular differentiation, and gives a
    perspective in which much of the work of the past decade can be in-
    terpreted.
J. Lash. 1974. In J. Lash and J.R. Whittaker, eds., *Concepts of Development.*
    Sinauer. A recent summary on the same topic with emphasis on car-
    tilage differentiation, mitosis versus differentiation, mass effects, etc.

Mammary gland development:
R.W. Turkington. 1972. In E. Litwack, ed., *Biochemical Actions of Hormones.*
    Academic Press. Vol. 2, p. 55. This paper leads to the literature on
    hormone effects on mammary development, DNA synthesis and milk
    proteins, and related topics.

Epidermal lability:
R.E. Billingham and W.K. Silvers. 1968. In R. Fleischmajer and R.E. Billing-
    ham, eds., *Epitheliomesenchymal Interactions.* Williams and Wilkins.
    This paper documents the remarkable dependence of even adult

42 epidermis on local dermal type. Other papers in the same volume comment on related topics in pancreas, muscle, cartilage, and other tissues.

Thyroid lability:

B.S. Spooner and S.R. Hilfer. 1971. *J. Cell Biol.*, *48*, 225. A demonstration of phenotype variations in relation to different nutrient mixtures.

Tissue stability in general:

H. Ursprung. 1968. *The Stability of the Differentiated State*. Springer-Verlag. A variety of papers presenting many points of view on stability or lack thereof.

Chapter Four:

The eye of a chick embryo that has been fractured to reveal the optic cup on the right, the lens vesicle in the center, and the early cornea covering the surface (left). Note that the hole into the lens seen in Figure 2.2 has been sealed over. (Courtesy of K.T. Tosney.)

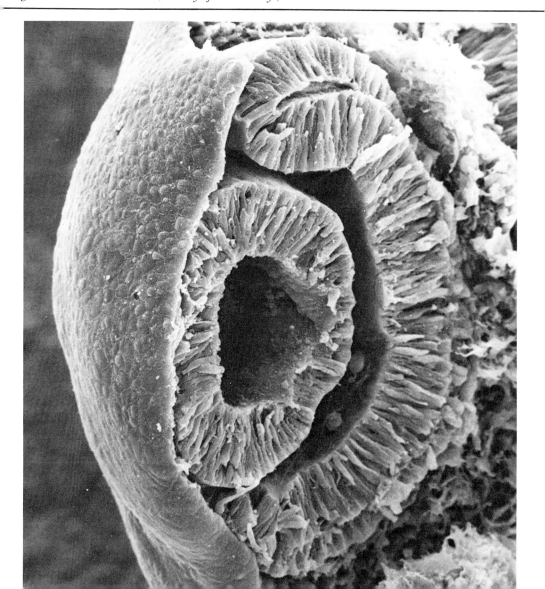

Types of Tissue Interactions

Is that an instruction, or just permission, said the ectoderm to the optic vesicle?

There are several types of interactions between the tissues of developing embryos. For example, if a portion of the early vertebrate eye, the optic vesicle, is placed in close proximity to head ectoderm cells that would, if left undisturbed, have formed epidermis or skin glands, then the ectoderm forms a lens instead (Figure 4.1). To do this, the responding ectodermal cells apparently employ genes responsible for lens protein production. Were it not for the presence of the optic vesicle, those genes would have remained or been placed in the inoperative, restricted condition, and keratin-, mucus-, or hair-producing genes would have been used by the same cells. In Holtzer's terminology, the optic vesicle is said to act *instructively* in causing this selective use of genes.

What does the experiment mean for normal lens formation? Two observations are pertinent (Figure 4.1): first, lenses form in normal

Figure 4.1

An idealized "cross section" cut through the head of a vertebrate embryo to illustrate an "instructive" tissue interaction. At region A, the optic vesicle and its direct descendant, the optic cup, have acted on the head ectoderm to cause lens formation. At B, the optic cup was removed before such action; as a result, the ectoderm comes under the influence of mesoderm cells, so that epidermis or hairs might form. No lens develops. At C, the optic vesicle removed from B was inserted beneath the ectoderm. There the mesoderm no longer influences the ectoderm, and a lens forms instead of epidermis or hair.

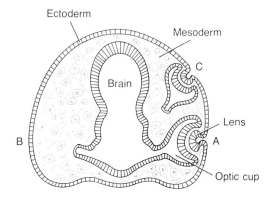

animals only at sites where optic vesicles approach close to head ectoderm; second, if optic vesicles are removed from the head, no lenses form. Therefore, three criteria can be specified for an *instructive* tissue interaction: (1) in the presence of tissue A, the responding tissue B develops in a specific way; (2) in the absence of A, tissue B fails to develop in that way; (3) in the presence of A, tissue C, normally destined to mature differently, is altered to develop like B, that is, in the specific manner associated with A. Of these criteria, the first two can also apply to permissive interactions (see below). Therefore, the third criterion is critical in defining an instructive interaction, since it is the only one that demands *control of selectivity in gene usage.* Even this type of action must be viewed as a necessary, but not a sufficient, cause of subsequent development—it is the rule, rather than the exception, that additional permissive interactions follow an instructive one in order that the fruits of the instructive one shall be realized.

There is one problem with the operations employed in establishing the third criterion for an instructive interaction that we must consider. It is that the potential responding tissue (the head ectoderm near C in Figure 4.1) must be "competent" to respond. If, for instance, we placed the optic vesicle beneath neck or flank ectoderm, then a lens might not develop. But this would be because the ectoderm in question has already come under other influences that restrict it; it then cannot employ the genetic and cellular machinery required to form a lens.

Thus the flank ectoderm might not be "competent" to respond to the optic vesicle. This does not mean that the optic vesicle is deficient in its instructive capabilities, but rather that the state of the responding cell system is a crucial factor in defining tissue interactions.

Two conclusions are warranted. The practical conclusion is that the experimental test for instructional capability must include carefully chosen responding tissues, perhaps from a variety of sources. The theoretical conclusion is that one must always be skeptical of negative results. As the brilliant American embryologist Ross G. Harrison pointed out, one can never know when a new set of conditions will allow the previously negative result to be reversed. Hence the test for an instructive interaction is at the same time a test for competency.

There is a fourth criterion for an instructive interaction that we should discuss. The responding tissue should not develop in the specific way in response to nonspecific stimuli. For instance, if head ectoderm formed lens when in proximity to pancreatic mesoderm, kidney epithelium, an ear placode, or even a piece of agar, then we could justifiably be skeptical that the same signal has emanated from those diverse sources to elicit the response. More likely, the ectoderm might be viewed as being in a "primed" condition, ready to go at the first opportunity. Obviously, this is a stringent criterion; furthermore, it is difficult to apply, since it is just imaginable that a variety of tissues might be able to send the same instructive signal. But that too seems to present logical difficulties, since it is hard to imagine why noneye tissues would emit molecules that effectively say "make lens." Hence we should keep this fourth criterion in mind, and see if and where it can be applied.

Now let us consider Holtzer's second type, *permissive* interactions. A simple example involves mitosis. A common observation is that mitosis ceases in embryonic epithelia (a piece of epidermis from skin, or the endodermal lining of the pancreas) placed in culture conditions (Figure 4.2). Similar epithelia cultured under identical conditions, but in the presence of embryonic mesodermal cells, maintain mitotic activity. Since a variety of mesoderm types (from lung, thyroid, kidney, etc.) can support mitosis in a given epithelium (e.g., pancreas), this permissive interaction is said to be of the "nonspecific" type.

What is important about this illustration is that the major expressive processes of tissue morphogenesis and cellular differentiation may

Figure 4.2
DNA synthesis by pieces of pancreatic epithelium cultured in the presence and absence of a mesodermal extract. Such synthesis is a measure of mitosis, and is actually ascertained by measuring quantities of radioactively labeled precursors of DNA bound in the cells. The decline in DNA synthesis in the presence of the extract that occurs after about 40 hours is related to differentiative processes to be discussed later. (After Ronzio and Rutter.)

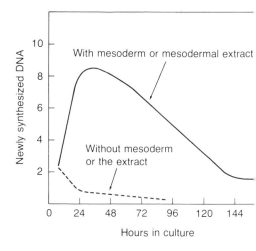

fail to occur in the nonmitotic epithelium cultured in the absence of mesoderm (lower curve in Figure 4.2). But both those processes will go on if mesoderm is added so that mitosis persists. We conclude that the permissive interaction between mesoderm and epithelium is *essential* for normal development. It is likely, however, that the action of mesoderm is primarily upon mitosis in epithelial cells, and only secondarily upon differentiation and morphogenesis. There is no obvious element of specific "information" content in this permissive interaction.

A second simple example of a permissive interaction operates in whole embryos. It concerns the Coulombres' observations on the chick eye. A long and complex series of tissue interactions, morphogenesis, and cell differentiation processes leads to the formation of the optic vesicle, the lens, and the cornea. Since these processes and their chemistry are normal, we would be inclined to conclude that all is well with development of the eye. That is too simple, however.

At about the fourth day of development of a chick embryo, the pressure of fluid within the main eye cavities begins to increase. This "intraocular pressure" plays a critical role in eye development. When the Coulombres inserted a tiny glass tube through the wall of the eye so that the internal pressure could not build up, many abnormalities appeared. An important one was that the cornea failed to acquire its normal curvature (see Figure 4.3); hence it could not bend light waves correctly to focus an image on the retina.

How did this come about? A ring of "scleral" bones develop around the outer edge of the chick cornea. The pressure within the eye apparently "pushes" outward against all the inner walls of the eye, including the inside of the cornea. The ring of bone seems to act like

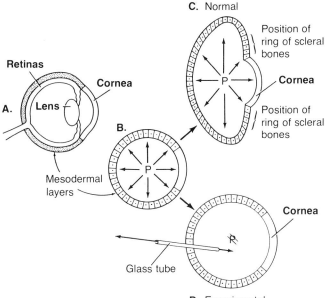

Figure 4.3

Effects of inserting a hollow tube through the wall of a developing eye. The outer mesodermal layers that give rise to the "sclera" and most of the cornea are shown in their normal relative positions in diagram A. For simplicity, only the sclera and mesodermal cornea are shown below. Normally, the increasing pressure "pushes" outward on all portions of the sphere (B). Because of the restraining effect of the ring of scleral bones, the cornea bulges as if it were a weak spot in the wall of a balloon (C). If, because of the tube, pressure fails to build up (D), the cornea fails to curve, the whole eye fails to expand to normal size, and, although they are not shown, grave abnormalities in retinal development occur. (After Coulombre and Coulombre.)

an inelastic restraining ring, which, because of its position, leads to the smaller radius of curvature of the normal cornea. Without the internal pressure, the process cannot go on. Thus both form and function of the cornea depend on intraocular pressure and, indirectly, on the cells responsible for that pressure.

This is a good example of a permissive interaction that does not depend on direct action of one cell type on another. Despite the normality of earlier tissue interactions, regulatory events, mitosis, and nucleic acid and protein syntheses, all is to no avail for the organism if the added permissive effect does not operate. Furthermore, this case shows that a physical parameter (the pressure) can have important effects on a developing system. It can act as a source of developmental information just as molecules can.

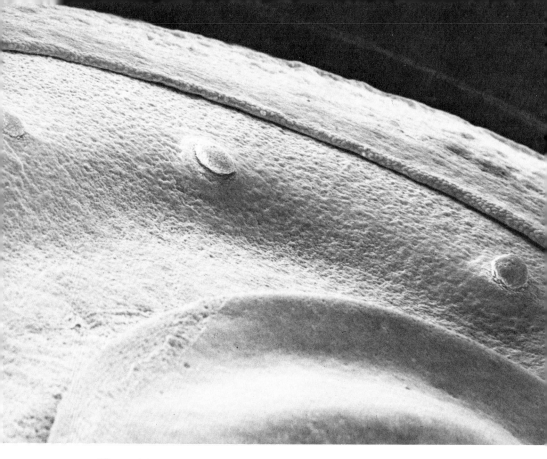

Figure 4.4

Lunar lanscape? No, the eye of a chick with the cornea below and three mound-like scleral papillae nearby. Fourteen of these structures develop in a circle around the cornea. Directly beneath each of them a scleral bone forms. When the Coulombres removed an early papilla, no bone formed beneath that site. Somehow these peculiar and transient accumulations of epidermal cells cause scleral bones to form. This interaction between epidermal cells and bone-forming mesodermal cells is an essential precursor to the intraocular pressure effects described in the text, since that phenomenon is dependent upon the ring of fourteen scleral bones. (See the frontispiece to Chapter 18 for a higher power view of a single papilla.)

Experimentation and Interpretation

It is appropriate now to pause and consider some of the experimental bases for our conclusions. Obviously, concepts such as permissive and instructive interactions do not arise from merely looking at embryos. Instead, embryos are manipulated; tissues are rearranged, removed to culture environments, or analyzed chemically. For the pancreatic epithelium and mesenchyme, the lack of mitosis has been demonstrated only in culture conditions (an epithelium cannot be effectively isolated from mesenchymal effects in an intact embryo). But any set of culture conditions is an arbitrary combination of nutrients, salts, temperature, etc. Therefore, conclusions must be couched in terms of the particular conditions employed. For example, experiments with differentiated thyroid epithelial cells show that mitosis can occur in the absence of mesoderm if a special nutrient mixture is present. Does this mean that a factor formerly supplied by mesoderm is now included as a nutrient? Does it mean that mitotic activity of other developing epithelia is a function of nutrients? Does it mean that these sorts of permissive reactions are artifacts of experimental design? There is no way of knowing at the moment.

Related to these questions are many types of experiments in which so-called "conditioned" media are used. Such media are usually prepared by culturing a large population of dissociated cells in a normal medium for several days. The medium is altered in largely undefined ways by exposure to the cells, so that if it is removed, resterilized, and used again, it may produce rather magical effects on another batch of embryonic cells. We shall encounter an example in Chapter 15 in which collagen and other substances in a conditioned medium will stimulate prospective muscle cells to fuse and develop into skeletal muscle fibers. Alternatively, cells from a nerve ganglion in the head (the ciliary ganglion near the eye) will not survive or form neurites (axons, dendrites) if cultured in very rich regular media, but live for weeks and form intricate neuritic networks if placed in a conditioned medium.

These results and many others to be encountered in the following chapters emphasize the point that a great deal of the experimental work on the early differentiation and morphogenesis of embryonic cells is, in Wilt's terms, "almost a history of the study of adequacy of media." The student should certainly not view all experiments on cell and organ culture with a jaundiced eye because of these reservations; nevertheless, constant awareness of the artificialities of the plastic and glass world of culturing must be kept in mind, so that appropriately conservative operational interpretations of results will be made. This is part of the dilemma faced by the experimental scientist: the experimental techniques used are essential if we are to advance knowledge, but those techniques can mislead if interpretation is not judicious.

We can now summarize our introductory discussion, which has set the framework for the more detailed discussions to come. Figure 4.5

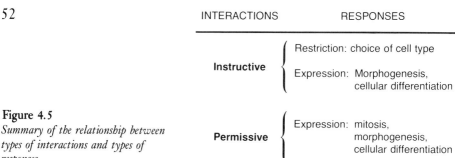

INTERACTIONS	RESPONSES
Instructive	Restriction: choice of cell type
	Expression: Morphogenesis, cellular differentiation
Permissive	Expression: mitosis, morphogenesis, cellular differentiation

Figure 4.5
Summary of the relationship between types of interactions and types of responses.

implies that only instructive interactions are involved when the choice of the determined state is being made. Permissive interactions do not act on the choice process, but instead on the basic mechanisms of organogenesis, mitosis, morphogenesis, and cell differentiation. In subsequent chapters we shall attempt to see if instructive interactions also act directly on the latter three processes. More importantly, we shall return to the question of whether the arguments presented in this chapter are valid. We will want to examine whether we are "forced" by experimental results to conclude that there is a real category of instructional interactions and a real class of informational molecules that are responsible for restriction and determination.

CONCEPTS

Instructive tissue interactions may affect the restriction-determination process, and so control which sets of genes are used during maturation of a cell line.

Competence is the ability of a cell population to respond to a developmental signal.

Permissive interactions do not affect the restriction-determination process, but instead are any of a variety of events that must occur if the expressive phase of cell development is to proceed normally.

The eye and determination:
A.J. Coulombre. 1965. In R.L. DeHaan and H. Ursprung, eds., *Organo-genesis*. Holt, Rinehart, and Winston. This is a summary of the instructive and permissive events in eye development, and it provides an excellent background to more recent literature.
G.V. Lopashov. 1963. *Developmental Mechanisms of Vertebrate Eye Rudiments.* Translated by J. Medawar. Macmillan. This book covers the huge literature on amphibian eye development, particularly the European school's work.

Determination in general:
R.G. Harrison. 1933. *Amer. Natur.*, 67, 306. One of the finest pieces of prose and of analytical thinking in all the literature of development. Any student of the subject should give this paper high priority in reading.

Mitosis and permissive interactions:
R. Ronzio and W.J. Rutter. 1973. *Develop. Biol.*, 30, 307. This paper provides the data on the factor extracted from mesenchyme that stimulates pancreatic epithelial mitosis.

Intraocular pressure and scleral bones:
A.J. Coulombre and J.L. Coulombre. 1961. In G.K. Smelser, ed., *The Structure of the Eye*. Academic Press. A review that includes discussion of the pressure phenomenon, as well as of transparency and other topics.

Thyroid cell phenotype and mitosis:
B.S. Spooner and S.R. Hilfer. 1971. *J. Cell Biol.*, 48, 22. An entry to the literature of Hilfer's group on the thyroid and its behavior in culture.

Chapter Five:

Feather germs growing from the back of a chick embryo. Each of these mounds will lengthen into a cylinder and then split into a frilled, light, down feather.

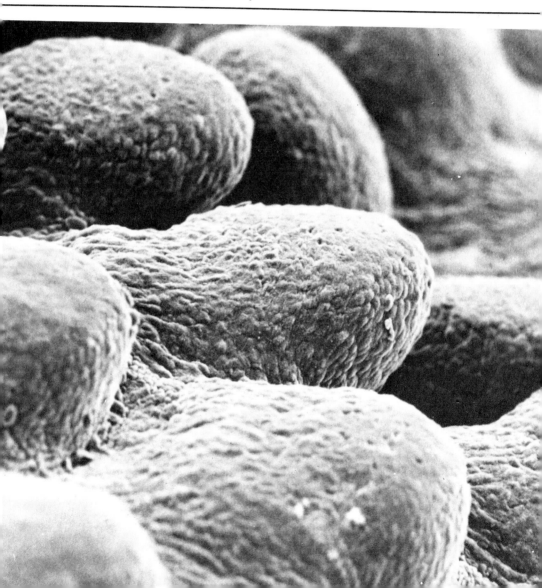

Skin and Its Derivatives

Chick cornea plus mouse dermis yields feathers? Why?

The development of vertebrate skin and its derivatives—hairs, feathers, scales, sweat glands, etc.—provides a classic example of mesodermal control of organ type. Skin itself is composed of two tissues, the epidermis, derived from ectoderm, and the dermis, derived from mesoderm. Feathers are exceedingly complicated in morphology and function. Consider the long, strong primary-flight feathers at a wing tip; the small, fluffy contour feathers on the breast; or the long, highly colored tail plumes of a male peacock that are used in display. Although all the diverse types of feathers are constructed almost completely of ectodermal cells, it is mesoderm that controls the type of feather that will form and much of its detailed morphology.

Three types of experiments performed by Rawles, Saunders, and Sengel demonstrate this (see Figure 5.1).

1. Saunders has shown that, when a block of mesoderm from the thigh of a chicken embryo is inserted beneath the ectoderm that covers the proximal portion of an embryonic wing, the wing ectoderm forms leg feathers. Both feather morphology itself and the arrangement of feathers on the wing surface are characteristic of the leg.

55

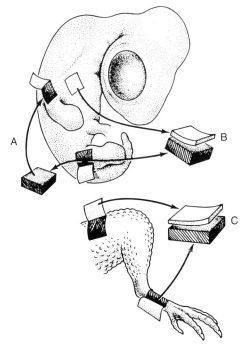

A. Leg mesoderm in wing → leg feathers
B. Feather-area mesoderm plus
   nonfeather-ectoderm → feathers
C. Scale-area mesoderm plus
   feather-area ectoderm → scales

Figure 5.1
*The action of specific types of mesoderm on ectoderm.
In every case the mesoderm controls the type of
derivative organ (feather, scale) that will form.*

2. Combination of feather-forming mesoderm with ectoderm located in
an area that does not form feathers results in feather development.

3. If a piece of ectoderm destined to form feathers is combined with mesoderm from the lower leg, where "scales" normally form, the ectoderm forms
scales. It has no intrinsic stability for feather formation.

All three of these results are consistent with the conclusion that
mesoderm controls the kind of specialized structure (scale or type of feather)
that will form from ectoderm.

In these experiments the mesoderm is actually calling into play
particular sets of ectodermal genes. This is shown by combining
mouse mesoderm (which would normally cause the mouse ectoderm
to form hair) with chick corneal epithelium (which would normally
become a curved, transparent surface). The result is *feather* formation!
The ectoderm has constructed the typical derivative organ of avian

skin. However, though "ordered" to grow an organ by mammalian dermis, the ectoderm can only respond by using its own genetic information. Hence, feathers develop, not hairs.

Philippe Sengel and his associates have performed many intriguing experiments in which skin tissues from different classes of vertebrates have been recombined. Reptiles, birds, and mammals have been used. Many modern reptiles are, of course, covered with scales, as were certain ancient reptiles, including the two distinctive groups that gave rise to the feathered birds and the hairy mammals. A primary finding is that a basic mesodermal signal for derivative formation operates in the skins of these diverse creatures. Both lizard and duck dermis would elicit hair formation from mammalian epidermis; human and lizard dermis would elicit feathers from chick epidermis; and so on. In all such situations, the derivatives proceed quite far in normal development. It is only during the terminal stages that abnormalities appear, reflecting an apparent inability of the foreign mesoderms to meet all of the epidermis' complex needs for full development.

Instead of working with different classes of vertebrates, suppose we limit ourselves to birds, and recombine chick and duck skin tissues. As would be expected from Rawles' and Saunders' results, discussed earlier in this chapter, in every instance the source of mesoderm controls gross feather structure. For instance, early duck feathers have a rigid rodlike backbone, the rachis, whereas early chick feathers do not. When duck mesoderm is combined with chick ectoderm, the feathers that form possess a rachis even though that structure is constructed entirely of chick cells. Only when one examines such feathers with great care can one see that the finest anatomical feature of feathers, the barbules (hooks that hold the barbs of the feather's vane together), are of typical chick shape and distribution (see Figure 5.2). It is only in these barbules that the genetics of the ectoderm cells are obviously expressed.

Apparently, duck and chick feathers are sufficiently alike that the foreign mesoderm can cause the altered arrangement in space of the chick cells which we recognize at a gross level as feather morphology. These results imply that there is a second type of mesodermal action during derivative formation. The first would be the general signal to "form a derivative." The second affects the actual expressive processes involved in alignment and arrangement of epidermal cells in order to generate organ structure; it affects processes of morphogenesis.

No doubt it is important for the two birds in this experiment to be close genetically. In the interclass recombinations described above, it seems likely that the relatively great genetic dissimilarity between

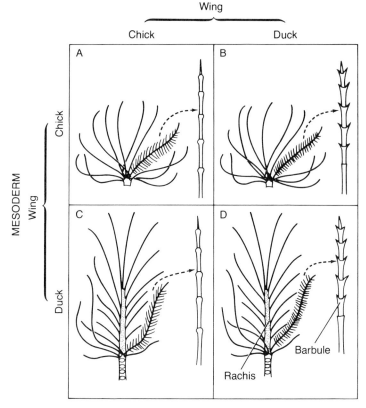

Figure 5.2

A chart summarizing results of recombining chick and duck wing ectoderms and mesoderms. **A** *and* **D** *show the controls of chick-chick and duck-duck. Note the absence of a rachis in* **A** *and its presence in* **D***. An individual barbule is drawn in the right portion of each panel; note the differences in morphology between* **A** *and* **D***. Now, observe in* **B** *that duck ectoderm does not form a rachis when in the presence of chick mesoderm, but its feathers do have "duck-like" barbules. Alternatively, in* **C** *it is seen that chick ectoderm forms a rachis when in the presence of duck mesoderm, although the barbules on the same feathers are of chick type. This reflects genetic limitations of the responding (ectodermal) system. (After* Advances in Morphogenesis, *9 (1971), 217.)*

birds and mammals precludes this second sort of signal from operating. Hence we see the derivative, but its development is aborted, presumably because the second kind of interaction cannot occur.

   Can we fit these experimental results into our scheme of restriction–expression and instructive–permissive interactions? Dermis certainly seems to cause populations of ectoderm cells to form organs that they would not normally form. This fact implies that there is an instructive interaction that affects restriction and determination. Furthermore, the remarkable effects of dermis on actual feather structure imply that a kind of "instruction" is affecting the expressive process or morphogenesis.

The reader should be aware, however, that a direct test of the restricted state of epidermal or derivative cells has not been attempted. One test might involve separation of the differentiating leg or wing-feather cells; activation of mitosis in them; and challenges of their capabilities for alternative types of development, like those applied to the cloned cartilage and pigment cells (see Figure 3.4).

Another direct test might involve precise analysis of the proteins manufactured when a single type of ectoderm is caused to form different types of feathers or scales by different mesoderms. Until such rigorous tests are completed, we have only the very strong inference that mesoderm controls restriction, but not a proof of that hypothesis.

Despite that very strong inference, it is *not* correct to make the further inference that mesoderm acts directly on the genetic material of ectodermal cells. The number of steps between combination of two

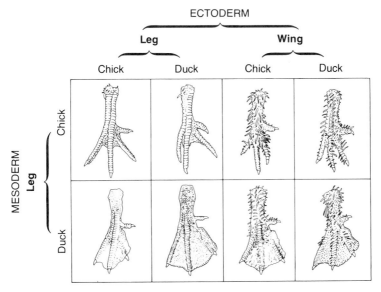

Figure 5.3
Webbing on a duck's foot is due to mesoderm. Control leg skin recombinations are seen in A (chick-chick) and D (duck-duck) respectively. Webbing between the toes also forms in C when duck mesoderm and chick ectoderm interact. In the four panels to the right (E-H), exactly the same rules operate for webbing. However, here wing ectoderm is used, and some degree of feather formation takes place on the developing scales. This reflects a tendency of the responding ectodermal tissue to employ genetic information appropriate for the region from which it came (wing, in this case). This is the same sort of behavior seen in Figure 5.2 for barbules, and it will be encountered in Chapter 6 for mesoderm in wings versus legs. (After Advances in Morphogenesis, *9 (1971), 217.)*

60 tissues and final organ development is undefined. In effect, there is a large black box between input (presence of mesoderm) and output (feather formation). We consider some of the features of that black box later in this chapter.

Permissive Interactions in Skin Development

We have seen that mesoderm acts instructively in feather development. It also acts permissively in several different ways. A piece of epidermis separated from mesoderm and cultured in a dish behaves just like the pancreatic epithelium of Figure 4.2: cell division and DNA synthesis virtually halt. Addition of mesoderm reactivates those processes. Like pancreas, an extract of embryonic tissue will also support DNA synthesis and mitosis in epidermis.

However, to act in this way, the epidermis must be in contact with a suitable physical substratum. Let us consider an example. Later in this book we shall discuss growth factors. One of these substances is a small protein that can be isolated from several sources and that

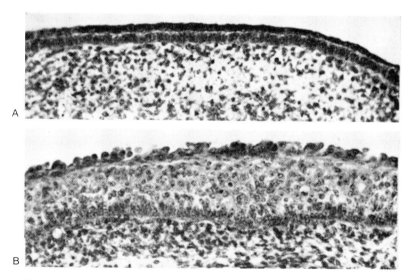

Figure 5.4

Back skin from chick embryos grown for three days in culture in the presence of control medium (A) or medium with EGF (B). The huge increase in epidermal thickness is caused by intense mitosis in response to EGF (as monitored by precursor incorporation into DNA). (Courtesy of S. Cohen. From Develop. Biol., 12 *(1965), 394.)*

Figure 5.5
A piece of embryonic chick epidermis cultured on top of a porous filter (F). The basal cells in contact with the filter are oriented and carrying out mitotic cycles (judged by DNA synthesis and visible mitoses). Those bridging the hole in the filter are flattened and do not incorporate precursors of DNA nor divide. Since the same nutrients are available to both sets of basal cells, the ability to respond to those substances, by carrying out mitotic cycles, must be governed by contact with the solid substratum.

stimulates mitotic activity in embryonic epidermis and other developing epithelia (it is called Epithelial Growth Factor or EGF). When Stanley Cohen removed chick epidermis from its dermis and cultured the epithelium on a porous filter to which it could become attached, then EGF exerted its action, and a huge number of cell divisions occurred. This worked only when the original lower surface of the epidermis contacted the filter. If the epidermis was inverted so that its upper (peridermal) surface was against the filter, then EGF could not stimulate DNA synthesis or cell division. Either the upper epidermal cells cannot attach effectively or they cannot respond to their attachment by becoming sensitive to EGF.

The same phenomenon is seen in Figure 5.5, where epidermis that is over a hole in a filter cannot respond to an extract of embryos even though bathed directly in it, whereas the epidermis that is in contact with the filter can do so. We can guess that in embryos the basal lamina, a sheet of extracellular macromolecules that is located at the interface of epidermis and dermis, acts like the filter and provides the necessary substratum for attachment.

Do these sorts of experimental observation have anything to do with the normal development of an embryo? Here is one fact that suggests they do. At a specific time, relatively early in epidermal development, the capacity to synthesize DNA and to divide becomes restricted to the lower basal cells. Cells situated above the basal layer, and *not in contact* with the basal lamina, cease to divide and begin to differentiate. A tentative explanation for this observation can be based on the experiments. Contact of epidermal cells with basal lamina maintains mitotic capacity and inhibits differentiation. Such cells retain competence to respond to EGF or other growth factors, and so the

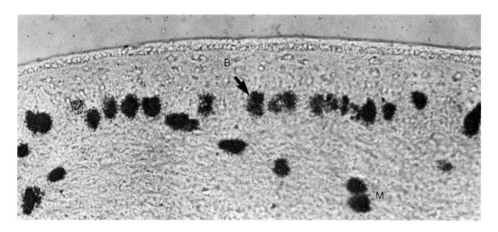

Figure 5.6

A piece of embryonic chick skin exposed to a radioactively labeled precursor of DNA. Only basal epidermal cells (B) are passing through the S phase of the cell cycle and so incorporate the precursor (some mesoderm cells are also in S). The radioactivity is visualized by covering the histological section with a photographic emulsion, exposing for several days in the dark, and then developing the emulsion, as one would photographic film. The exposed silver grains in the emulsion appear black, indicating where the precursor is localized. If the same experiment had been performed with younger skin, then cells located at all levels of the epidermis would have incorporated the precursor; i.e., they would all have been carrying out mitotic cycles.

normal thickness of epidermis can be regulated. Once cells move outward, away from contact with the basal lamina and dermis, they lose that competence, cease dividing, and differentiate.

We conclude that both the growth factors and the attachment to a physical substratum are permissive factors for mitosis and normal tissue development. Both are essential, but neither is sufficient by itself.

The development of feathers also provides an illustration of a permissive factor that may be nonchemical in character. Sengel (1976) noted that, when the neural tube–notochord system was removed from a portion of an early chick embryo, no feathers would form later in the back skin in the region of the embryo that had been operated on. He also noted that insertion of various dense objects (ranging from pieces of agar or chemically inert nitrocellulose filters, to masses of certain tissues) beneath the skin of an embryo in regions where feathers never normally form is followed by feather development in the overlying skin. It seems likely that the inserted objects act mechanically in some way to cause dermis immediately beneath the epidermis to become quite densely packed. This appears to be the essential precondition for all the complicated later events of feather development. It is 63 noteworthy that neither the mesodermal nor the ectodermal cells in that area of the embryo would have used feather-specific genes were it not for this "mechanical" action. This is a good lesson that very complicated chains of developmental events can be set in motion by seemingly simple events. Sengel's experiments may be interpreted to mean that the neural tube and somites of the normal embryo also act mechanically in causing dermis to become dense so that feathers develop (other evidence we shall not discuss raises the possibility of chemical interaction between the neural tube and dermis). This speculative interpretation implies that essential biological "information" can be a physical force

Figure 5.7
The sizes and shapes of individual cells are highly variable on different parts of these early chick-feather germs and on the intervening epidermis. It is not known what causes such differences in cell shape, nor whether the shapes are significant for feather and skin development.

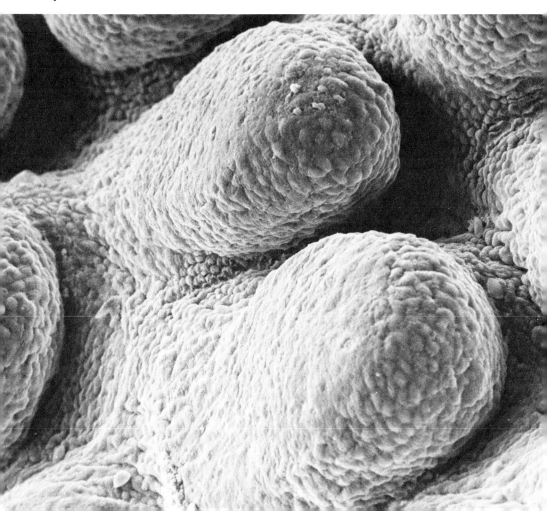

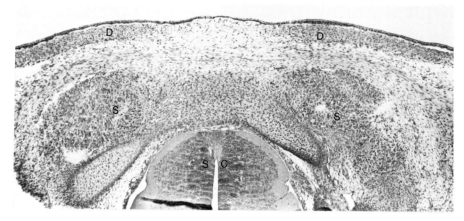

Figure 5.8

A cross section through the back of a chick embryo. The spinal cord (SC) and upper portions of the somites (S) are seen. Overlying the somite regions are areas of dense dermis (D), whereas directly above the spinal cord the dermis is less dense, as is characteristic of more primitive dermis. Feather germs will develop only in the region of dense dermis. It is not clear whether a chemical signal from the spinal cord leads to densification of the dermis and feather formation, or whether a "mechanical" effect from the underlying tissues is sufficient to initiate the chain of events that culminates in feather morphogenesis.

(the mechanical derangement resulting from presence of a piece of filter material or the neural tube), not just specific molecules.

Though the hypothesized "mechanical" action may help to initiate feather development on the back of an embryo, many other processes are essential in the actual formation of the organs and in the propagation of the pattern of feathers over the surface of the skin. Deposition and orientation of extracellular substances such as collagen, movement of cells from the somites to the skin, mitosis of cells, and other expressive events are interwoven to yield the final development of organs arranged in a precise way. Thus a given event in the series, such as the mechanical action, must be viewed as an essential, but not a sufficient, cause of the over-all process.

CONCEPTS

The type of mesoderm (dermis) that is present controls the local characteristics of epidermis and its derivatives (feathers, hairs, etc.).

The basic signal emanating from dermis to stimulate derivative formation is relatively nonspecific; it acts across class lines in vertebrates.

Among closely related organisms, dermis can control aspects of the shape of epidermal derivatives. This implies an instructive interaction acting on the expressive process of morphogenesis.

An important permissive effect in skin is attachment of epidermal basal cells to a suitable substratum.

Either systemic or local factors may control mitosis in epidermal basal-cell populations.

REFERENCES

Skin and feathers:
P. Sengel. 1976. *Morphogenesis of Skin.* Cambridge Univ. Press. This superb book covers all the important experiments and theories on skin development, is a model of analytical thinking and writing, and cites all significant papers through 1973 (including ones on duck-chick recombinants).
D. Dhouailly. 1975. *W. Roux Arch. Develop. Biol., 177,* 323. This is the latest treatment of reptile, bird, mammal recombination experiments.
M.E. Rawles. 1963. *J. Embryol. Exptl. Morphol., 11,* 765. This is one of the classic papers by the founder of this field.

Feather specificity and mesoderm:
J.W. Saunders and M.T. Gasseling. 1957. *J. Exptl. Zool., 135,* 503.

Thigh mesoderm interaction with wing ectoderm:
J.W. Saunders and M.T. Gasseling. 1959. *Develop. Biol., 1,* 281. These two papers are the primary sources for the respective interactions.

Epidermal Growth Factor:
S. Cohen. 1971. In M. Hamburgh and E.J.W. Barrington, eds., *Hormones in Development.* Appleton-Century-Crofts. *Develop. Biol., 12* (1965), 394. *J. Investig. Dermatol., 40* (1972), 1. These papers summarize the early work on EGF, from both the biological and the biochemical points of view.

Substratum requirement for mitosis:
N.K. Wessells. 1964. *Proc. Natl. Acad. Sci., 52,* 252. Experiments on substrata and nutrients in relation to mitosis.

Genetic mutants, the collagen lattice, and feather patterns:
P.F. Goetinck, and M.J. Sekellick. 1972. *Develop. Biol., 28,* 636. An easy entry to the literature on this topic, as well as a good illustration of how tissue recombinants can be used on mutant tissues.

Chapter Six:

The apical ectodermal ridge of a chick wing bud viewed here from the front of the left wing. The dorsal surface of the wing is to the right of this AER and the ventral surface to the left and below. This "ridge" of cells has remarkable developmental properties. (Courtesy of K.T. Tosney).

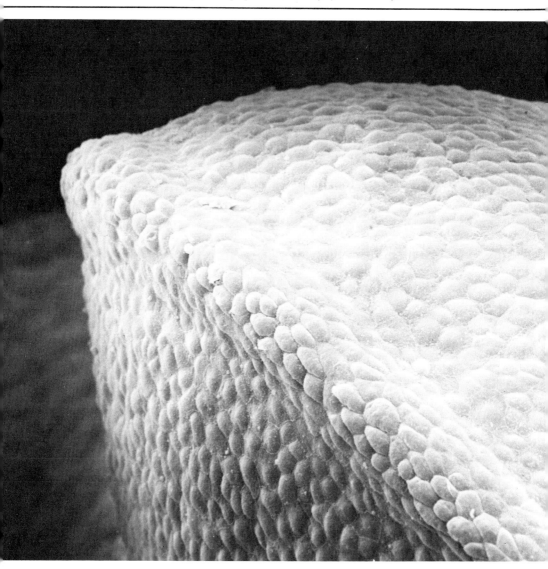

Limbs

Archaeopteryx, the earliest bird, had toes on its wings!
Mix a little thigh and a little wing, and we could have the
same on that chicken embryo.

Experiments on limbs provide some of the best evidence that two tissues may affect one another in important ways during organ formation. If hind-limb mesoderm is transplanted under forelimb ectoderm, the result is outgrowth of a hind limb. Just as with skin derivatives, it is the mesoderm that controls the type of limb development. This is not surprising, since, unlike feathers, the limb is composed mainly of mesodermal tissues (muscles, bones, tendons, etc.). Contributions of the ectoderm to limb development are very important, however, as is indicated by the following observations.

Soon after limb outgrowth starts, a thickened ridge of ectodermal cells caps the distal portion of the limb. This "apical ectodermal ridge" (AER) is essential for limb outgrowth. If it is removed at any time, limb outgrowth ceases (Figure 6.1, A), without, however, any adverse effects on regions closer to the base of the limb. Both mesodermal (bones, muscles, etc.) and ectodermal (feathers, epidermis) structures

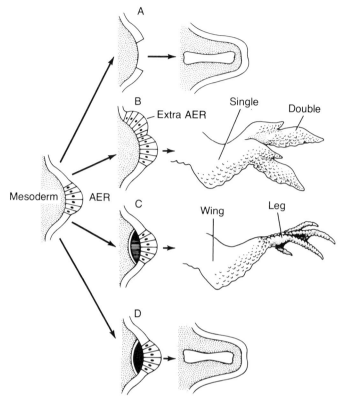

Figure 6.1
*A summary of relationships between the AER and underlying
mesenchyme. The diagrammatic section of a young limb bud shows the
apical ectodermal ridge (AER). If the ridge is removed (**A**), outgrowth
ceases but development in the stump continues. If an extra AER is
inserted (**B**), the distal limb is doubled. If leg mesoderm is placed beneath
the AER (**C**), structures appropriate for the hind limb develop from
that point outward. Finally, if nonlimb mesoderm is grafted beneath the
AER (**D**), the AER regresses and outgrowth stops.*

of the proximal limb continue to develop normally. If the apical ectodermal ridge is not removed, but an extra one is grafted onto a limb
above or below the original ridge, an additional axis of outgrowth is
created (Figure 6.1, B). From that point onward, the distal portions of
the limb are doubled. It is concluded that the ridge serves as an orienting influence for outgrowth.

The environment in the immediate vicinity of the apical ectodermal ridge has remarkable properties. Recall the experiment (see
Figure 5.1) in which a block of thigh mesoderm caused thigh feathers
to form from proximal wing ectoderm. If an *identical* piece of thigh

mesoderm is grafted beneath the apical ridge of a wing, the surprising result is that toes and distal leg parts develop on the end of the wing (Figure 6.1, C). The thigh mesoderm cells that would have participated in formation of thigh tissues and organs (dermis, feathers, blood vessels, and perhaps muscles) instead form toe bones, tendons, muscle, and blood vessels. In addition, such former "thigh" mesoderm acts back on the distal wing ectoderm to cause it to develop scales (plate-like epidermal thickenings that cover the distal leg and toes). Thus not only is the mesoderm cell's own pattern of development altered, but also its capacity to serve as a source of information in an apparent instructive interaction is changed. Proximity to the AER produces these transformations in the mesoderm. Something about the ridge environment ensures that cells nearby shall behave appropriately for distal parts of a limb.

Figure 6.2
A fractured chick limb bud showing the AER and underlying mesoderm cells. Note that the ridge cells are much more elongated than the normal ectoderm cells that cover the rest of the limb bud. (The sphere beneath the AER is a preparation artifact.) (Courtesy of K.T. Tosney.)

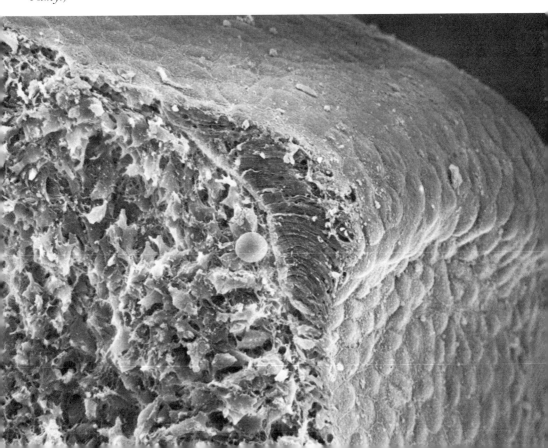

70   How do these results relate to the concept of restriction–determination? We know that thigh mesoderm cells would never normally use the combination of genes required to generate toes and the distal portion of a hind limb. Since they can do so when influenced by the environment of the ectodermal ridge, they cannot by our definition be fully restricted in developmental capability (recall that the determined state is *heritable*). However, they *are* probably restricted to forming structures of the hind-limb type. The *wing* apical ridge cannot cause the thigh cells to develop as distal wing cells; the thigh cells develop only as distal hind limb cells and form toes. In this respect they are like chick ectoderm responding to mouse mesoderm (see Chapter 5). Chick feathers form presumably because only that kind of species-specific genetic information is available for use. With thigh mesoderm, only one species or even one organism need be involved; nevertheless, the cells are limited to forming organs of *hind-limb* type. What better illustration could be constructed to elucidate what is meant by restriction? In fact, such thigh mesoderm provides a clear example of restriction but not final determination—lability remains until the cells participate in final organ development.

The importance of the initial restriction to "limbness" was seen by the embryologist Edgar Zwilling when he placed a piece of non-limb mesoderm beneath the apical ridge (Figure 6.1, D). For instance, if mesoderm from the back (the somites) is put under the ridge, the ridge regresses, and limb outgrowth halts. This happens even though somite mesoderm also has the capacity to differentiate into muscle, bone, cartilage, and connective-tissue cells that would be essentially indistinguishable from ones in a limb. It is reasoned that limb mesoderm is the sole source of an apical ridge "maintenance factor" that is essential to the function of the ridge. Whether the "factor" is a specific molecule or a capacity to interact in some special way with ectodermal ridge, or whether it involves some other process or mechanism, is unknown. We shall return to this maintenance factor when we discuss certain mutations that alter limb morphology.

The Progress Zone

Why do limb-cell populations develop the correct proximo-distal sequence of bones, muscles, and other tissues? The AER and immediately adjacent mesoderm has been termed a "progress zone" by Wolpert, Lewis, and Summerbell. The zone is believed to be the site

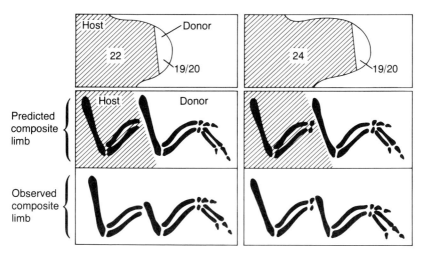

Figure 6.3
Grafts of AER-mes from young (stage 19/20) donor limb buds to older host limb buds. The predicted outcome from progress zone theory is indicated, and typical results are shown. (From Summerbell.)

where "assignments" are made to cell populations so that they will develop in specific ways. It is hypothesized that a "positional value" is assigned that subsequently insures formation of humerus, or ulna-radius, and so forth.

How do these hypotheses arise? First, imagine a graft of an AER and the immediately adjacent mesoderm (we will abbreviate the combination as "AER-mes") from a young wing bud to an older one whose tip has been removed (see Figure 6.3). Development proceeds and the host stump region forms a humerus. The thin area of mesoderm on the implant proliferates and yields a series of bony elements arranged in a correct proximo-distal sequence: humerus, ulna-radius, wrist, phalanges (see Figure 6.4). Most important is the presence of two humeri (arranged in sequence).

How can this be explained? We hypothesize that the mesoderm of the host stump already had its positional value assigned (i.e., to form humerus) so that the population of cells continued the process of development. The implant, however, behaved independently, and proceeded to assign a correct series of positional values, just as it would have done if left undisturbed in its original site.

Note that the stump does not instruct the AER-mes system to form the next-most distal skeletal segment (ulna-radius would be appropriate in our example). Instead, the AER-mes behaves quite independently, as if no communication concerning positional values travels in a proximo-distal direction.

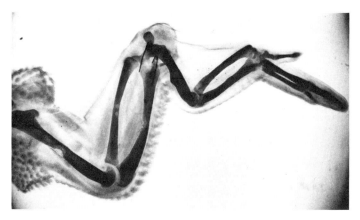

Figure 6.4

A photograph of a composite wing arising from the kind of graft diagrammed in Figure 6.3. The host stump has developed a normal humerus, radius, and ulna. The added AER-mes has formed its own humerus, radius-ulna, wrist, and digit parts. (Courtesy of J.H. Lewis, D. Summerbell, and the Journal of Embryology and Experimental Morphology.*)*

What if we do the reverse experiment? Let us graft the tip of an "old" wing bud on the stump of a younger one (see Figure 6.5). We observe that intermediate wing segments are missing! For instance, the wrist and phalanges might be located just beyond a proximal half-humerus.

How can this be accounted for? Assume our operation on the young host happened to cut through the young wing-bud mesoderm midway out in the population of cells whose positional value was "humerus." The result is the half-humerus we observed. Now consider the implant. Since we put on the AER-mes of an "old" wing bud, positional values for humerus and ulna-radius already would have

Figure 6.5

A tracing of a wing that results from the experiment opposite to that in Figures 6.3 and 6.4. Here the AER-mes from the "old" limb bud was grafted on a "young" host limb bud. A humerus has formed from host tissue. No radius-ulna segment is present. The hand (phalanges) developed from the old implant. Therefore, we see no hint that the proximal host tissue influences the peripheral AER-mes; if it had, then radius-ulna and wrist should be present. (Courtesy of J.H. Lewis, D. Summerbell, and L. Wolpert. From Nature, 244 *(1973), 492.)*

been assigned. Hence, when grafted, the AER-mes continues to behave normally from its point of view, and assigns the next segment, wrist. There is no intervening ulna-radius in the composite wing. Again, there is no evidence of proximal to distal flow of information. If there were, the missing bones would be present.

Quantal Nature of Positional Assignments

Saunders found that AER's from limbs of differing ages could be exchanged without effect on limb development. Since this is so, the sub-AER mesoderm *must* be the site where positional values change and are assigned. How does this region function? Summerbell removed the AER of wings at different stages (as in Figure 6.6), and noted that one always gets complete skeletal segments that are appropriate for the time when the operation was performed: that is, a *complete* humerus, or complete humerus and ulna-radius, or completeness out

Figure 6.6
*Chick limbs that developed after removal of the AER at stage 19 (**A**) or stage 20 (**B**). Development through the humerus is complete in **A** (the two bones seen are the scapula of the shoulder girdle and the humerus). The humerus, radius, and ulna have developed in **B** (the scapula is present at the left). If the AER had been removed at a later stage, then wrist or digit parts would have developed too. (Courtesy of D. Summerbell and the* Journal of Embryology and Experimental Morphology.)

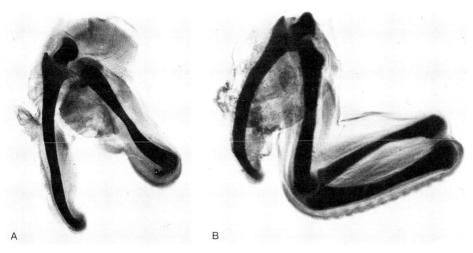

A   B

74   through the wrist. One rarely sees half-humeri or three-quarter ulnas. It seems, then, that positional values are assigned in quantal steps corresponding to the major segments of the limb. It is as if the AER-mes is in a "make humerus" state for a specific time, then "make ulna-radius," then "make wrist," etc. There is no hint that the AER-mes system operates as a continuously varying system whereby a sequential set of precise assignments are made within each skeletal region (e.g., make proximal one-eighth ulna, next eighth, etc.).

Interestingly, the time spent by the AER-mes in each of its states corresponds to the time that elapses during a single cell-division cycle (about eight hours). Of course, cell divisions in the AER-mes are asynchronous and continuous, and cells do not stop dividing when their positional value is assigned and they leave the AER environment. Instead, the AER-mes shifts through its seven states (humerus, ulna-radius, carpels I, carpels II, metacarpels, phalanges I, phalanges II), taking about eight hours for each. In the Summerbell-Lewis terminology, the AER-mes behaves like "a clock whose ticks are cell-division cycles."

We said that precise assignments of proximo-distal position within each of the seven skeletal segments are not made when cells leave the AER-mes. That type of information arises because of interactions within each of the populations of dividing cells whose over-all positional value has been defined. This can be demonstrated in the following kind of experiment (see Figure 6.7). Let us take the core of the prospective humerus region from the center of a limb bud, mark the cells with radioisotopes or some other means that will let us recognize them, and then implant them at right angles to the axis of the forming humerus in a host limb bud. Do this so that one half of our marked population protrudes laterally into the area destined to form muscle next to the humerus. Development proceeds, and we find that marked cells in the humerus region have participated in forming bone. But the marked cells located more laterally form muscle and connective tissue, just as is appropriate for the environment in which they reside. Clearly those cells did not have their final determined state fixed when their positional value was originally allocated.

An important component in this choice process between muscle and bone may be the blood vessels within the limb mesenchyme. The distributions and relative sizes of various blood vessels may influence oxygen, carbon dioxide, nutrient, and hormone levels within the core of the limb bud. These factors, in turn, could affect the choice between muscle and skeletal (bone, cartilage) differentiation. We know very

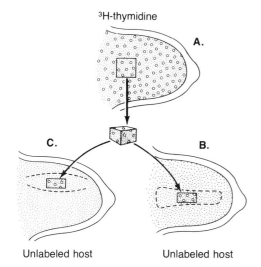

³H-thymidine

A.

C.

B.

Unlabeled host   Unlabeled host

Figure 6.7
*Experiment showing effects of local environment on cartilage formation. First (**A**), a developing limb is exposed to ³H-thymidine, a precursor of DNA, long enough for most cells to be labeled. Then, blocks of prospective cartilage-forming cells are removed and implanted in unlabeled host limbs. If placed near the core of the host limb bud, the labeled cells participate with host cells in forming cartilage and bone (**B**). If implanted more peripherally, the implanted cells participate in forming muscle or dermis (**C**). The site controls the fate of the implanted cells. (After R.L. Searls and M.Y. Janners, J. Exptl. Zool., 170 (1969), 365.)*

little about the control of blood vessel development, though in the case of amphibian limb regeneration (Chapter 11) nerves may be essential for blood vessel formation. The roles and significance of blood vessels in development of limbs and of other organs are worthy of intense scrutiny.

These experiments show that the relative positions of bones and muscles, the shapes of bones, and similar features are worked out at the local level in the limb. In fact, these sorts of observations emphasize that, as far as we know, positional values are assigned to cell populations; we do not understand yet how such information operates for single cells.

Now that we understand the functioning of the AER-mes system, we can reconsider Saunder's experiment of placing prospective thigh mesoderm beneath a wing AER. If the progress-zone hypothesis is correct, it follows that positional values are not fixed with the same degree of stability as "legness" or "wingness." To the contrary, the thigh tissue comes under the influence of wing AER (and nearby wing mesoderm), and proceeds to assign appropriate distal values, in the correct sequence. In a sense, the "clock" can be set ticking again in a cell population whose positional value had been assigned. This implies strongly that positional values are not heritable in the way that early states of restriction ("legness") or determination are. One can only wonder whether this means that positional information operates at the cell-population level and is not a stable property of individual cells.

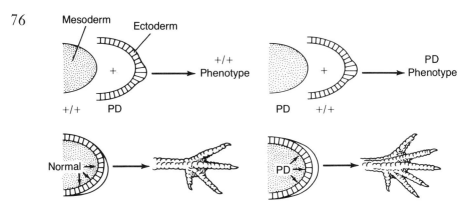

Figure 6.8
Diagrammatic representation of combining polydactylous tissues (PD) with normal, wild-type tissues (+/+). Mutant ectoderm shows no evidence of expression of the mutation. The hypothesis offered to explain the mutation is also diagrammed, the limb buds being viewed from their dorsal surface so that the greater relative size of the AER when PD mesenchyme is present can be seen.

Limb Mutants

The significance of the AER maintenance factor is illustrated by certain limb mutants in birds. These mutants are also of interest because they serve as models to explain the sort of developmental abnormality seen in occasional embryos of all terrestrial vertebrates, including humans—namely, extra, missing, or fused digits, abnormally shaped limbs, and the like.

"Polydactyly," meaning "extra digits," is a dominant, nonlethal mutation found in a number of organisms. In domestic chickens, where this anomaly has been studied, mutant embryos form extra toes next to normal ones on the feet. Developing limb tissues from mutant embryos can be recombined with limb tissues of normal embryos to test whether the genetic defect is a property of ectoderm, mesoderm, or both. Normal mesoderm plus mutant ectoderm yields normal numbers of toes (Figure 6.8). Mutant mesoderm plus normal ectoderm produces extra toes. Therefore, the mesoderm is the site where the mutation exerts its effect. Observations of developing polydactylous mutant limbs reveal that the apical ectodermal ridge is abnormally broad: instead of being restricted to the tip of the limb, the site where toes normally form, the ridge extends anteriorly and posteriorly farther than normal, and in such regions extra toes appear. Since the genetic effect is operant in mesoderm, it has been proposed that the mutation causes an abnormally widespread distribution of apical-ridge maintenance factor. Overlying ectoderm responds by forming the extra wide ridge, the ridge acts back on mesoderm to orient outgrowth, and the outcome is extra toes.

Precisely the same relationships are seen in human polydactylous conditions. An extra-wide AER, extending far more anteriorly or posteriorly than normal, is associated with the subsequent production of extra fingers or toes. It will be intriguing in future years to learn whether the same gene products are altered in birds and mammals to lead to the polydactylous syndrome.

"Eudiplopodia" is another mutation in domestic fowl, in which an extra plate of toes forms above the normal one (as if you had five extra toes protruding from your shin). Recombinations of mutant and normal tissues similar to the ones described above demonstrate that the ectoderm is the site of expression of the mutation. Mutant epidermis plus normal mesoderm yields the extra set of toes; the reverse combination gives the normal number. When developing mutant limbs are examined, a lengthy normal course of development results in a proximal limb region that shows no hint of the mutation. Then, an extra epidermal ridge appears in a position dorsal to the normal one (Figure 6.9). The extra ridge serves as a secondary outgrowth site, just as experimentally implanted ridges do (see Figure 6.1). The secondary site undergoes development of an extra set of toes.

These mutants establish that genetic lesions in either of two interacting tissues can have marked consequences for development. We do not know what proteins are produced by these or the many other alleles that function during limb development. Therefore, we cannot guess about the mechanisms affected when such genes are mutated. But the mutants teach us an important lesson: there are probably large numbers of alleles functioning in the normal development of all organs that we cannot recognize because we do not have mutations of them, or know what their protein end products are.

The limb mutants are also of considerable general interest to biologists, since one of the most important series of evolutionary changes in terrestrial vertebrates has involved modification of limb structure to allow the use of diverse ecological niches that is characteristic of those organisms. Examples include the alterations of forelimbs into the distinctive wings of flying reptiles (pterosaurs), bats, and birds, or the well-documented evolution of the limbs of horses wherein the number and size of toes have varied. The very means by which such evolutionary modification can occur is seen in the developmental limb mutants.

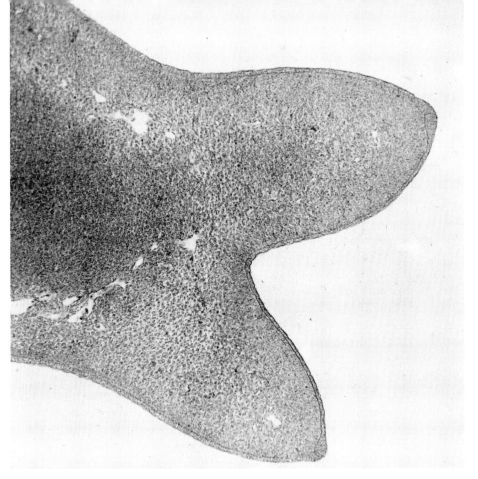

Figure 6.9
A section through the limb bud of a eudiplopodia mutant embryo is seen here. The normal axis of outgrowth is toward the bottom of this photograph. An extra apical ridge system has developed on the dorsal surface of the limb bud and protrudes toward the upper right. This tissue will give rise to the extra plate of toes characteristic of legs in these mutants. (Courtesy of P. Goetinck. From Current Topics in Developmental Biology 1, *(1966), 272.)*

Axes and Orientation of Limbs

If an arm or leg or wing is to function optimally in an adult, it must be oriented correctly relative to the main body, and its various parts must bend and move in specific directions as well. The orientation of your hand and arm did not occur by chance. If some skillful experimenter had rotated one of your limb buds at an appropriate time, your palm might face upward, your thumb point backward. How do the correct axes of symmetry arise?

A "zone of polarizing activity" (ZPA) has been investigated by Saunders and his colleagues (see Figure 6.10). If one removes a block of mesoderm from near the posterior junction of an early limb bud and

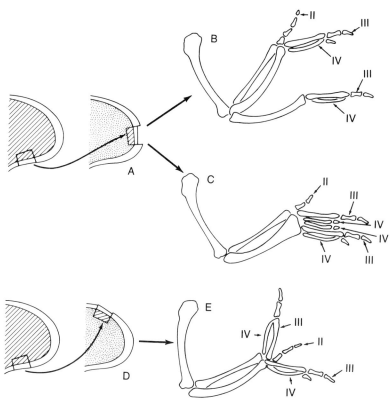

Figure 6.10

*ZPA transplantation experiments. In **A** the ZPA region is placed near the tip of a host limb. Outgrowth may take several forms, two of which are shown. In **B**, the anterior hand is normal (digits: II, III, IV); whereas the posterior part of **B** is deficient anteriorly (digits: II missing; III, IV present). In **C**, a single complex outgrowth has formed with digit order: II, III, IV, IV, III, IV. If the ZPA region is placed anteriorly in a host limb (**D**), the anterior outgrowth (**E**) has fully reversed polarity, and digit II may be shared! (Digit order: IV, III, II, III, IV). (After A.B. MacCabe* et al., Mechs. of Aging and Devel., 2 (1975), 1.)

the lateral body wall, and implants that block beneath the ectoderm at the front of a host wing bud, then a duplication develops. An extra wing grows out from near the implant site and the orientation of the wing is reversed. Its *posterior* side is toward the implanted ZPA (i.e., the posterior side of the wing is nearest the head of the host chick embryo). If a ZPA is grafted beneath an AER, duplication occurs from that time onward. The posterior side of the extra, anteriorly situated duplication is located nearest the implanted ZPA. More interesting is the fact that the host limb's distal outgrowth shows varying degrees of reorientation: frequently it is a virtual double-posterior (its own

80   original posterior faces its own ZPA; the front half of the same bones, being situated near the implanted ZPA, also develop like "posterior" structures). We conclude, therefore, that a ZPA has two capacities: it can stimulate limb-bud outgrowth; and it governs anterior-posterior axes. For a limb, "posterior" equals "towards the ZPA."

A "wingless" mutation has been known in chickens for some time. In homozygotic embryos no wings develop and only abortive leg buds form. If the region of wing mesoderm located where a ZPA should be found is taken from a homozygote embryo and combined with wild-type ectoderm and AER, no wing forms. Apparently, the wingless mutation eliminates the ZPA, and so no wings are initiated. This kind of observation implies that normal limb-bud initiation may occur because of action by the ZPA. We do not know how this kind of stimulatory activity relates to control of the anterior-posterior axis, though one might guess that independent mechanisms are involved.

The dorso-ventral axis of a limb becomes fixed at a different time in embryonic development than the anterior-posterior axis. There are some hints that ectoderm plays a role in controlling dorsality and ventrality. For instance, one can remove the mesodermal core of a limb, dissociate it to single cells, pack the cells together again, and return them to their ectodermal jacket. When placed on a host embryo, the implant grows into a limb. The limb lacks any hint of an anterior or posterior side; it is symmetrical in that respect. This means that the dispersion of the ZPA region cells and random mixing among other limb mesoderm cells effectively eliminates the ZPA information that controls anterior-posteriorness.

However, the implanted limb does have a top and bottom. From the direction toes bend, from feather and scale distributions, we can identify dorsal and ventral surfaces; and these surfaces always coincide with the original dorsal and ventral sides of the ectodermal jacket.

Extracellular Factors and Wing Development

Three experiments raise the possibility that soluble or diffusible factors are involved in AER or ZPA functions. First, if one places the mesoderm removed from just beneath an AER (the mesoderm from a progress zone) in the bottom of a culture dish with a fluid medium present, then many of the cells begin to die after about 10 hours. However, we can place about 6 AERs in a tiny basket suspended in the fluid medium, and then find that the mesoderm cells do not begin dying at

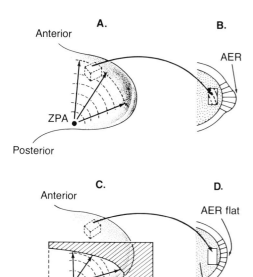

A. Anterior / ZPA / Posterior

B. AER

C. Anterior / Posterior / Mylar

D. AER flat / Mesenchyme dies

Figure 6.11
ZPA activity and barrier effects. In a normal limb (A), the ZPA region is located near the posterior edge of limb mesenchyme. If a block of mesenchyme is removed from the anterior half of such a limb and implanted near a host AER (B), the AER is maintained, and underlying host mesenchyme remains healthy. If an impermeable barrier is inserted in the limb for 24 hours before the same transplant is done (C), then the host AER (D) flattens, and many underlying host mesenchyme cells become necrotic. (After MacCabe and Parker.)

10 hours. Some factor from the AERs facilitates survival. (As a control, pieces of the same mesoderm can be cultured, still in contact with dorsal wing-bud ectoderm, but *without* any AER itself. Such mesoderm begins to die at 10 hours, just as if no ectoderm was present. It seems, therefore, that the capacity to aid the mesoderm cells is restricted to the AER itself.) We do not know yet whether the factor from AER that sustains mesoderm cells in culture is the means by which the AER orients wing outgrowth.

In a second type of experiment, we culture an AER-mes system in fluid medium. After a few hours, the AER ectoderm cells begin to flatten, a sure sign that their function is impeded. Now, suppose we add 6 to 8 pieces of posterior limb-bud mesoderm in a basket to the culture medium. Such tissue is, of course, the source of the putative maintenance factor and the site of the ZPA. The result is that the AER does not flatten. Despite being separated from the responding tissue by a distance of about 1.5 millimeters, the AER maintenance factor of posterior mesoderm can function!

The final experiment involves the ZPA effect. Using the cultured AER-mes system described above as an assay system, McCabe and Calemba have shown that a macromolecule or macromolecular complex can be isolated from the posterior limb mesenchyme and added to a culture medium, where it causes the AER to remain thick (functional). In an intact limb bud, this substance may originate from the

82   ZPA region and diffuse forward through the core of the mesenchyme. This is implied by an experiment in which an impermeable sheet of Mylar, a plastic, is inserted into a wing bud to isolate the anterior and posterior halves (see Figure 6.11). If mesenchyme from the anterior region is assayed for ZPA-factor activity immediately after the barrier is inserted, then the AER remains thick. However, a similar assay conducted several hours after the Mylar barrier is inserted shows an absence of ZPA-factor activity. One might imagine, therefore, that the ZPA serves as a source of a substance that passes forward through the mesenchymal spaces and that acts on the overlying ectoderm to preserve the AER. Presumably, the material is degraded or inactivated in time, thus accounting for its gradual disappearance from isolated anterior mesenchyme. Only future work will tell us whether this macromolecule, and its distribution, is the vehicle by which the ZPA controls the anterior-posterior axis of a limb, or whether the substance is equivalent to Zwilling's maintenance factor.

To conclude, it should be evident that we are approaching the time when the biochemistry of higher-order phenomena, such as axes of orientation, can be investigated. With the further development of unambiguous assay systems, perfection of isolation and purification methodology, and careful interpretation, even these seemingly magical aspects of an embryo may be explained in mechanistic terms.

Instructive or Permissive Interactions in Limbs?

Clearly, limbs develop because of a complicated series of interactions between ectodermal and mesodermal tissues. Limb mesoderm may initiate outgrowth, support AER function, and is also the site of progress-zone function. The AER, in turn, serves as a point of outgrowth, is essential for maintenance of progress-zone activity, and may govern dorsal-ventral symmetry.

The function of an AER does not fit the definition of an instructive interaction that was proposed earlier in this book. It is much more difficult to decide about the diverse roles of limb mesoderm. On the one hand, AER formation and function may be viewed as a primitive kind of differentiation; one might guess that specific genes are involved when the AER functions. Furthermore, limb mesoderm grafted beneath flank ectoderm can cause an AER to form and a limb to grow forth. In a formal sense, the combined action of ZPA and maintenance factor meets all four criteria for an instructive interaction (the student should go through the reasoning that leads to this conclusion).

Nevertheless, our concepts are clearly being strained when we try to apply them to this complex situation. That strain may reflect defects in the theoretical framework, as well as a basic ignorance of the chemical mechanism that must underlie these tissue interactions. Consequently, the more cautious conclusion is that a series of essential, but permissive, interactions go on during limb development. The information leading to assignment of positional values, to pathways of differentiation, to axes of symmetry, and to directions of outgrowth, may emerge from properties of the mesodermal and ectodermal populations. Only future work will tell us whether signals that are involved in the choices of gene restriction and determination need be exchanged between the limb tissues.

CONCEPTS

Mesoderm controls the type of limb that forms.

Limb mesoderm cells pass through a stepwise restriction process, one step of which can be characterized by the terms "forelimbness" or "hindlimbness."

The AER serves to orient limb outgrowth.

The "progress zone" consists of AER plus immediately adjacent limb mesoderm.

The progress zone is a site where "positional values" are assumed by mesodermal cell populations.

The assumption of positional values may occur in a "quantal" manner, the steps corresponding to each of the major skeletal segments in a limb.

The "zone of polarizing activity," a region of posterior limb mesoderm, governs the anterior-posterior axis of a limb.

The ectoderm governs the dorsal-ventral axis of a limb.

Limb mutants and evolutionary variations in vertebrate limb structure can be understood in terms of known properties of AER, the progress zone, the ZPA, the maintenance factor, and other variables that influence limb development.

84   REFERENCES

The progress zone:
D. Summerbell and J.H. Lewis. 1975. *J. Embryol. Exptl. Morphol.: 33*, 419 and
621; *35* (1976), 241. These papers summarize the current position of
the Wolpert group on the progress zone; included are references to
earlier work as well as the "quantal" nature of assigning positional
information.

J.W. Saunders *et al.* 1976. *Develop. Biol.*, *50*, 16. This and an adjacent paper
in the journal consider AER function and point out difficulties with
the progress zone hypothesis.

"Limbness" and early restriction in the limb:
E. Zwilling. 1968. In M. Locke, ed., *The Emergence of Order in Developing Systems*. Academic Press. An evenly balanced discussion of conflicting
hypotheses on limb development, including the maintenance factor
and "limbness."

Mutants and the limb:
P.F. Goetinck. 1966. In A.A. Moscona and A. Monroy, eds., *Current Topics in Developmental Biology*. This review describes some of the chicken
mutants that have been used most effectively in analyzing limb and
skin development.

Zone of polarizing activity:
J.A. MacCabe *et al.* 1973. *Develop. Biol.*, *31*, 323. This paper discusses anterior-
posterior and dorso-ventral axes and their control.

Diffusible factors and limbs:
J. Cairns. 1976. *J. Embryol. Exptl. Morphol.*, *34*, 155.

J.A. MacCabe and B.W. Parker. 1975. *Develop. Biol.*, *45*, 349. These papers
cover work on cultured AER and the sub-AER mesoderm in which
soluble factors operate. A further paper from MacCabe is in press,
Develop. Biol., 1976.

Chapter Seven:

The right half of a mouse embryonic lung from which much of the investing mesoderm has been removed. The stage of this specimen approximates that seen in the right portion of the "36-hour" drawing in Figure 3.1. As the single bronchial buds protruding toward the upper right in this photograph continue to grow in size, they will bifurcate and form complex branched systems. (Courtesy of E. H-L. Hu.)

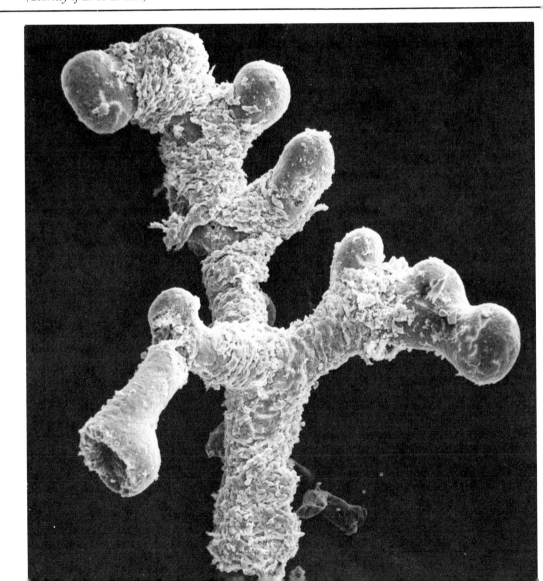

Tissue Interactions
and Morphogenesis

*Cells elongate, become narrow, or move about, and so
generate the changing form of an embryo.*

Much of our knowledge about the control of morphogenesis comes
from studies that deal with the internal organs of vertebrates. The
lungs, salivary glands, and kidneys, for example, contain two dissimilar cell populations. One population consists of epithelial cells
that line the elaborate ducts and functional units in these organs. The
other consists of mesenchymal cells that ultimately make up the tough
"connective" tissue and the blood-vessel system of the organs. (The
reader should note that we are *not* contrasting "endodermal" with
"mesodermal" cell types here, because some organs, such as the kidney, are wholly of mesodermal origin. What appears to be significant
is the "epithelial" tissue arrangement, in a cohesive sheet, versus the
"mesenchymal" arrangement, in a jumbled mass of cells; see Figure 7.1).

What are the interactions between the two population types that
lead to morphogenesis in such organs? Answers are suggested by
experiments in which epithelial and mesenchymal rudiments are
manipulated in culture. If the epithelium of a mouse embryonic lung
is cultured under the best nutrient conditions we can devise, it becomes

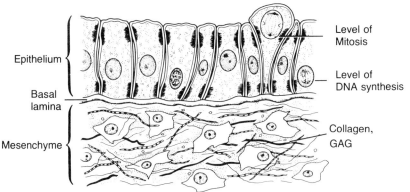

Figure 7.1

Epithelium and mesenchyme. An epithelium is a population of cells bordering a free space. Characteristically, regions of junctional complexes at the outer and inner lateral cell surfaces hold the cells in a cohesive sheet. The nucleus can be displaced in the cytoplasm, and is usually located near the inner end of the cell during the S period (DNA synthesis), near the outer end during mitosis itself (M). A basal lamina is located at the interface of the epithelium with the mesenchyme. The latter tissue is a looser aggregate of cells scattered amidst collagen fibers and other extracellular materials. The density of packing of embryonic mesenchymal cells varies greatly both between tissues and with time.

cystic and fails to branch out in a tree-like fashion (as it would have in the embryo), even when the substances needed to support mitosis are supplied (see Figure 4.2). Thus, high levels of embryonic extracts allow mitosis, but do not support morphogenesis. On the other hand, if the lung epithelium is recombined with lung mesenchyme before it is cultured, then the epithelium branches profusely.

Alternatively, let us scrape away mesoderm from a small area next to the epithelium of the trachea (the tube that leads to the lung bronchial system). When we insert a piece of lung bronchial mesoderm into the gap and culture the tissues for a few hours, an extra lung bud grows forth (Figure 7.2). The region of trachea that responds in this manner never would have participated in any sort of branching morphogenesis, if left undisturbed.

A third type of experiment establishes a different property of some mesenchyme. Suppose we take the tracheal mesoderm that was removed in the experiment just described, and place that tracheal mesoderm around the bronchial portion of the lung epithelium. Branching halts, and the localized area of bronchial epithelium that is in contact with tracheal mesoderm does not undergo further morphogenesis, whereas all the surrounding portions proceed to branch normally (see Figure 7.3). Clearly, there is a difference between tracheal and bronchial mesoderms—the latter allows (stimulates?) branching, the former does not allow (inhibits?) branching. So, even in a single organ,

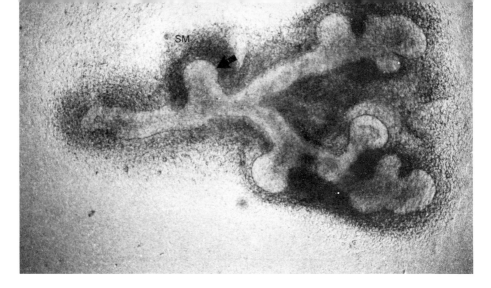

Figure 7.2

An embryonic mouse lung system. An area of tracheal mesoderm was removed, and a piece of mouse embryonic salivary gland mesenchyme (SM) was inserted next to the trachea. Here, a day later in culture, a bud (arrow) has grown outward from the side of the trachea. This experiment shows that the ability to stimulate the initial bud morphogenesis is not limited to the mesoderm that normally surrounds bronchial buds; it is even present in mesoderm from another organ.

Figure 7.3

The converse of the experiment in Figure 7.2. Here, tracheal mesoderm (TM) was placed over the tip of the left primary bronchus (LB) of a mouse embryonic lung epithelium. Two days later, as seen here, the normal right bronchus (RB) has branched, whereas morphogenetic branching in the left bronchus is completely inhibited because of the presence of tracheal mesoderm. The trachea (T) is on the right.

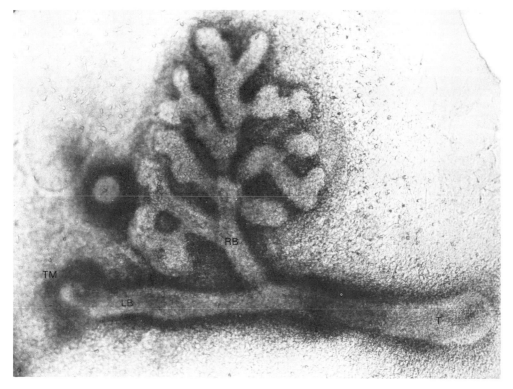

90   there may be subpopulations of mesodermal cells that have different capabilities in tissue interactions.

Branching morphogenesis of other internal organs (salivary, kidney, etc.) is also believed to depend on local mesenchymal populations. There is still uncertainty about how specific these sorts of tissue interactions are. Lawson has made the important discovery that salivary-gland epithelium will branch extensively when it is in the presence of large quantities of *lung* mesenchyme (or of smaller quantities that are not allowed to spread away from the epithelium).

The crucial nature of experimental design is shown by Lawson's observation that salivary morphogenesis will not occur if smaller quantities of lung mesenchyme are employed (or if they spread away from the epithelium). In fact, earlier workers had repeatedly observed this absence of morphogenesis and had erroneously concluded that only salivary mesenchyme could meet the needs of the salivary epithelium. Lawson's results, and similar results on the kidney, argue that we need not assume great specificity in the interaction that supports morphogenesis. It seems that most of the requisite information is resident in the epithelial cell populations.

However, as with most things in biology, there are exceptions. For instance, in the salivary epithelium and lung mesenchyme recombinations, the initial epithelial lobules that formed were rather broad and flat, thereby resembling early stages of lung epithelial morphogenesis. An even more impressive case comes from Kratochwil, who placed mouse mammary epithelium within salivary gland mesenchyme; the resultant lobules were unusually broad and densely packed; they resembled salivary epithelium much more than the open, narrow duct system of a mammary gland (see Figure 7.4).

In both of these examples, morphogenesis of the epithelium is altered, and the epithelium resembles that of the organ from which the mesenchyme is derived. A type of instructive interaction appears to occur, analogous to that we encountered for skin derivatives. Recall how duck dermis can alter the morphology of chick feathers. However, all the epithelia we have discussed undergo substantial morphogenesis in the presence of foreign mesoderms. It seems prudent to conclude, therefore, that the basic interaction is permissive. Thus, we may regard the epithelial populations as being restricted or determined, and able to carry out the appropriate expressive process—morphogenesis in these examples—if permissive requirements are met. We shall consider the means by which the various mesenchymes may alter epithelial morphogenesis, after we look at the mechanisms which underlie that process.

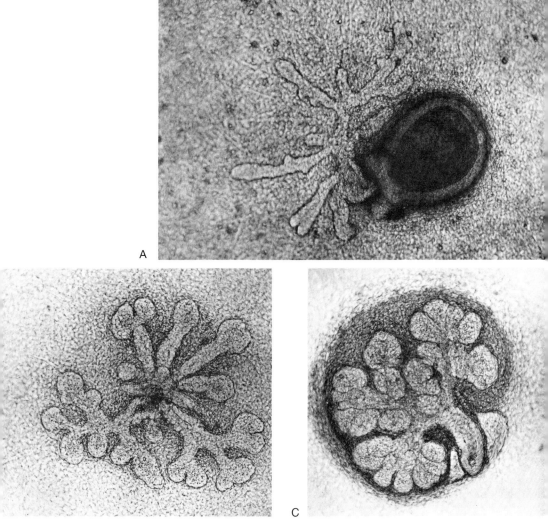

Figure 7.4

Kratochwil's experiments with mammary and salivary tissues. In A, mammary epithelium cultured with its own mesenchyme forms narrow, tubular structures. In B, when cultured with salivary mesenchyme, it forms more bulbous endings with cleft-like slits at the ends, and bears a distinct resemblance to salivary epithelium in its own mesenchyme (C). The remarkable behavior seen in B certainly implies that salivary mesenchyme can act "instructively" on the morphogenetic process of mammary epithelium. (Courtesy of K. Kratochwil. B and C are from Develop. Biol., 20 (1969), 46.)

Cell Behavior during Morphogenesis

Four basic processes can account for morphogenesis in multicellular animal embryos: (1) localized cell division, (2) localized cell death, (3) localized deposition of extracellular matrix, and (4) cell movements. The last category may involve single-cell migration or movements of cell populations. Let us describe briefly an example of each process.

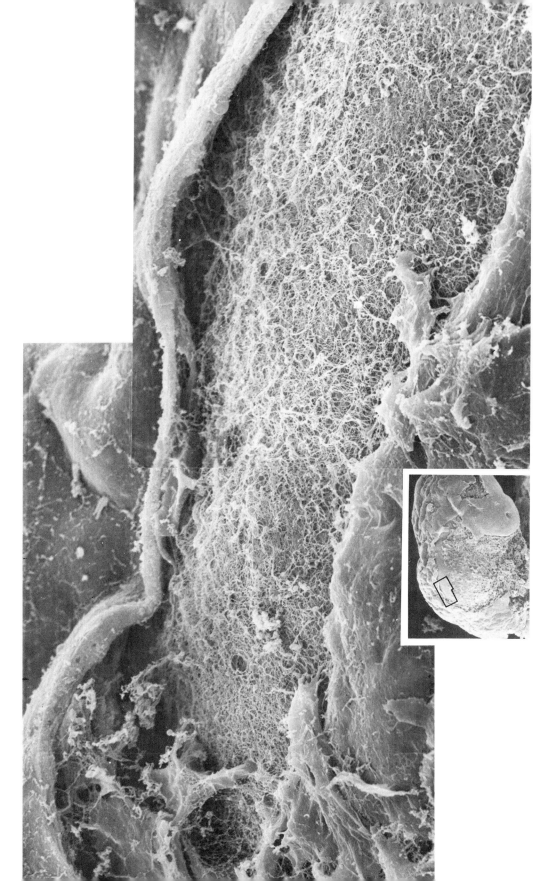

(1) At any early stage in feather germ development, mitotic cells are distributed at random throughout the mesenchymal area of the germ. Then, as outgrowth from the surface of the embryo becomes noticeable, few mitoses are seen in the anterior half of the germ, but many occur posteriorly. Excess daughter cells accumulate posteriorly, and the elongating germ grows out pointing toward the tail. Here the asymmetric distribution of dividing cells influences the orientation of the organ relative to the embryonic body.

(2) Mesenchyme cells located in the peripheral part of the limb bud die if they happen to be situated in the regions between developing digits. This leads to separation of the digits; abnormalities in this "normal" cell death lead to webbing between the fingers or toes. Here cell death is responsible for sculpturing of the limb.

(3) Large amounts of proteins, protein-sugar complexes, and insoluble salts of calcium and phosphate accumulate between the cells of developing bone. The cells are separated from each other, and a substantial volume is ultimately occupied by the bony matrix. Here extracellularly situated materials are shaped to yield bones of characteristic size and shape.

(4) Single cells of populations such as the neural crest (a mass of cells that arises at the top of the closing neural tube in vertebrate embryos) migrate for relatively long distances over rather precise routes before they settle down and develop into a variety of differentiated cell types (see Figure 7.5). Alternatively, sheets of epithelial cells fold in various ways to generate complex tree-like structures such as the lining of the lungs. In these examples, single cells or coordinated groups of epithelial cells are carrying out movements that contribute to morphogenesis.

Morphogenesis in an Epithelium

To fully appreciate the complexity of morphogenesis, we shall turn to several different developing organs in order to see how these four basic processes are integrated to yield the final form of organs. First,

◀ Figure 7.5
The substratum for cell locomotion is seen in this view of the head of a chick embryo. The head is seen in the small inset photograph. The eye with the hole leading into the primitive lens primordium is at the lower right. Head ectoderm has been stripped away and the area outlined in black is enlarged in the main photograph. Neural crest cells migrate over and through this tangled web of collagen fibers upon which sugar-protein complexes have been precipitated. (Courtesy of K.T. Tosney.)

94   the salivary gland of mouse embryos will provide an excellent example of morphogenesis that is based on the coordinated activity of a cell population.

The "branching" morphogenesis of salivary epithelium proceeds as follows: the outer surface of each rounded lobe is a site where a narrow cleft appears. These clefts gradually widen; the adjacent portions of the lobe expand in size; and they, in turn, are bifurcated by new clefts. A tree-like structure emerges (see Figure 7.6). In kidney or lung epithelia, somewhat different shapes are seen but analogous processes may go on (see Figure 3.1).

Localized differences in mitotic rate may play a role in some types of morphogenesis. In salivary glands, after discrete lobes have appeared and are separated from each other by deepened and broadened clefts, cells that are synthesizing DNA (in the S phase) before mitosis are seen more frequently at the tips of the lobes than near the bases of the clefts. This distribution can account for the continuing peripheral expansion of the lobes beyond the region where each cleft forms. It is dangerous to extrapolate from one developing organ to another, however, because extensive observation of lung epithelium (as in Figure 3.1) show a random distribution of DNA-synthesizing cells and little hint of localized concentrations of mitoses at the bronchial branch points. Consequently, we must assume that each organ may utilize local differences in mitotic rates to varying degrees during morphogenetic branching.

Do cells in an epithelial sheet, such as those in the salivary epithelium, move about and change their nearest neighbors during morphogenesis? In the only example in which this question has been examined carefully, cells in the amphibian embryonic medullary plate may change shape drastically, but they do not change position relative to their neighbors. Thus, we might visualize a morphogenetic epithelium as a sheet of rubber, capable of being deformed, but not composed of small pieces constantly milling about each other.

Though epithelial cells may not alter their relative positions, they should not be pictured as static bricks from which the new structure is built. In fact, we have been severely misled by thinking about developing tissues, particularly epithelia, on the basis of what is observed in fixed and sectioned specimens, whether by light or by electron microscopy. When a developing salivary gland is photographed by time-lapse cinematography, the epithelial surface is seen to be highly dynamic, moving in and out, forming numerous transient clefts, and almost bubbling with activity. One gains the impression that cleft after

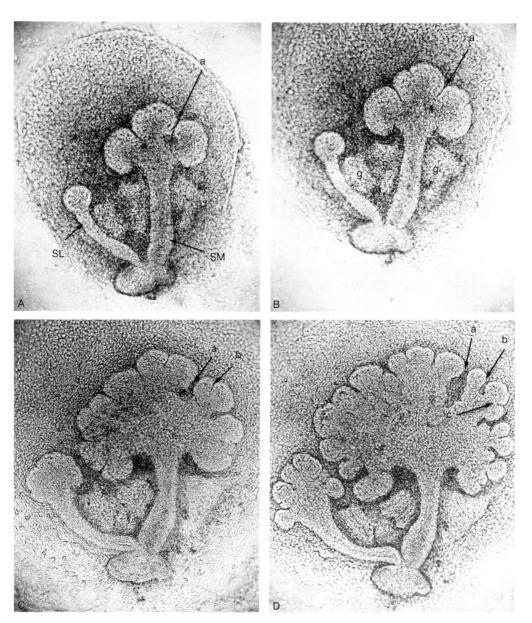

Figure 7.6
Four stages in the morphogenesis of a salivary gland in culture. Both the submandibular (SM) and sublingual (SL) glands are present. Note how cleft A widens; then cleft B appears, and subsequently widens (photograph D); by that time, cleft C has appeared nearby. The same kind of process can be seen elsewhere on the photographs. This process of cleft formation and epithelial growth continues as a more and more complexly shaped epithelial "tree" is generated. Observe how the same kind of morphogenetic process is carried out by the sublingual epithelium. The structures marked G are parasympathetic nerve ganglia; many axons grow from these ganglia over the surface of the two branching epithelia (see Figure 11.8).

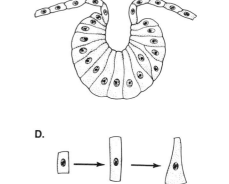

Figure 7.7
Diagrammatic representation of changes in cell shape during neural tube formation in an amphibian embryo. The somewhat elongated ectodermal cells (A) lengthen in the region of the medullary plate, whereas those located laterally flatten as they form epidermis. The medullary plate then rolls into a tube (C); as this occurs, the highly elongate cells of the medullary plate narrow at their outer ends and become flask- or wedge-shaped. The stages seen in any one cell are diagrammed below (D). (Redrawn from B. Burnside, Develop. Biol., 26 (1971), 416.)

cleft starts, then one "makes it" and goes ahead to deepen, while others regress, to form again elsewhere. We have not even begun to include this kind of "real life" phenomenon in our hypotheses about morphogenesis, though it would not be surprising that analogous dynamic movement is the rule, not the exception, during epithelial morphogenesis.

Morphogenesis in epithelia almost always includes two types of gross change in shape. First, individual cells may elongate (see Figure 7.7). This process is the earliest recognized expressive event during development of lens, feathers, thyroid, pancreas, and many other organs. (Why this is so is an interesting but unanswered question.) The second change in cell shape is a narrowing at one end of individual cells. The lateral cell surfaces at that end approach one another, apparently with sufficient speed that the nearby end surface of the cell is thrown into fingerlike processes or sheets.

How are such changes in shape of individual cells related to morphogenesis in a sheet of cells? The answer to this question must be understood if the reader is to appreciate what is meant by *integration* in development of multicellular organisms. Imagine that a discrete subpopulation of cells in our epithelial sheet enters a state wherein the inner ends of the cells begin to narrow. The cells are firmly anchored to one another along their lateral surfaces. The narrowing process results in a substantial reduction in the surface area at one end of the subpopulation (Figure 7.7). Only one result is possible: the cell *population* changes shape; i.e., the flat sheet forms a groove or depression. What has happened is that a change occurring within single cells is translated into a morphogenetic movement of the cell population as a whole. Note that the grooves do *not* appear in individual cells; grooves are a characteristic of the cell population.

If the change in cell shape is the basis of morphogenesis in an epithelial sheet, then the means of maintaining or altering cell shape are of crucial importance. Cell elongation is associated with the presence of cytoplasmic microtubules—long, seemingly rigid filaments that are aligned parallel to the axis of cell lengthening. If cells are treated with colchicine, an alkaloid compound isolated from certain plants, cytoplasmic microtubules are dispersed and cell elongation cannot occur. It is not known yet whether the microtubules act as rigid, structural rods that generate directly the forces for cell elongation, or whether they serve as tracts along which "flow" of cytoplasm occurs. Either process could be the basis for cell lengthening.

The narrowing of cells may be attributable to another type of intracellular organelle: bundles of microfilaments. Filaments in such bundles are inserted on the inner side of the lateral plasma membranes in specialized regions called junctional complexes (see Figures

Figure 7.8
Microtubules (M) in an elongating axon of an embryonic nerve cell. These rigid-appearing structures may act as bonelike skeletal struts, as railroad tracks to transport cytoplasmic components, or in both these ways. Microtubules similar to these frequently run parallel to the long axis of cells that are carrying out elongation or participating in morphogenetic movements.

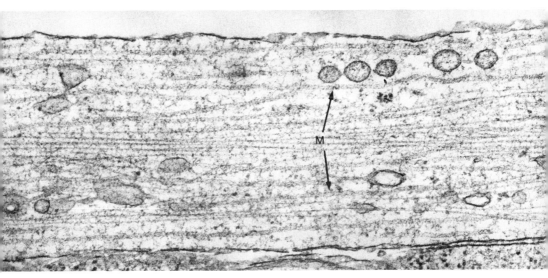

98   10.4 and 13.2). It has been proposed that the microfilaments are a primitive kind of contractile system. (Microfilaments of similar arrangement act as contractile agents in other systems. In addition, they bind a portion of the contractile protein, myosin, derived from muscle cells; this is a strong indication that the microfilaments are very like "actin" in structure.) If microfilaments are contractile, then it is intriguing to consider the means by which they might be controlled. Smooth muscle, the most primitive type of the three major classes of

Figure 7.9
Microfilaments (F) associated with the lateral surfaces of mouse embryonic pancreas cells. The bundles of microfilaments may insert into the dense, black-appearing materials that are associated with the plasma membranes of these cells. Contraction of the bundles could then exert force upon the cell surface and thereby contribute to change in or maintenance of cell shape. (T: thick "tonofilaments," whose function is unknown.)

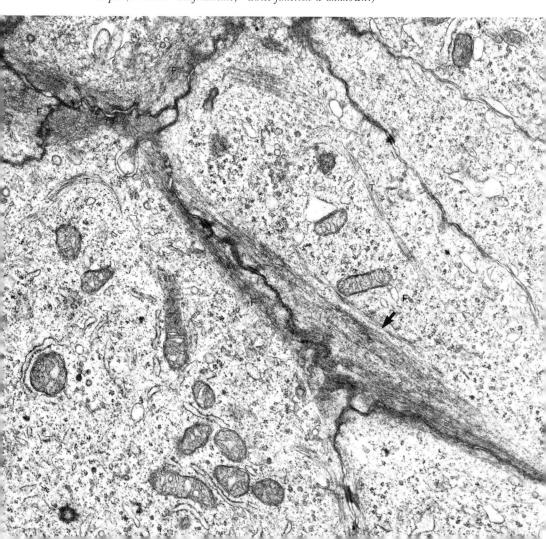

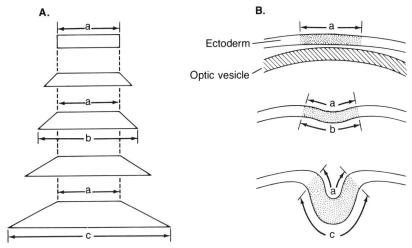

Figure 7.10
Possible results of having one "stabilized" surface on a growing epithelium. (A) Imagine that dimension a is fixed; as cell growth and division go on, the epithelium may get thicker or wider on its other surface (dimensions b, c). (B) As an example, suppose contact between optic vesicle and head ectoderm leads to the same sort of "stabilization" of the outer ends of the head ectoderm cells. Subsequent growth and mitosis generates an increase in volume and mass that causes the lens population to bulge inward, forming the lens vesicle. In other cases, an interaction might lead to stabilization of the surface near the second tissue, and the responding population would bulge outward. This hypothesis does not require active contraction of microfilament bundles for generation of the change in shape. (See Zwann and Hendrix.)

muscle in vertebrates, depends on calcium ions and ATP for contraction. Experiments with microfilament systems in cleaving embryos have also shown calcium-initiated contractions. Several types of experiments have therefore been performed on morphogenetically active thyroid and salivary epithelia in which either a so-called "contraction medium" (containing appropriate levels of ATP and metal ions) or drugs that inhibit calcium permeability have been used. The results are consistent in implying that these epithelia behave like smooth muscle. However, there are such minute quantities of tissue involved that direct measurements have not yet been made of the putative contractile tensions that are produced in the cells located at critical sites near morphogenetic folding.

A simple alternative to the idea that changes in the shape of individual cells *generate* folding of epithelial sheets comes from observations by Zwann on the lens. The infolding of the sheet of head ectoderm to form the lens vesicle might result from a simple process. Imagine that in the region of contact between the optic vesicle and the head ectoderm, a condition is set up which affects the outer surface of the ectoderm. That is, the ectodermal cells are altered in some way, so that

Figure 7.11

An early stage in lens development in a chick embryo. The two-layered optic cup is on the left. The thick lens primordium on the right has sunk inward from the surface of the head, and remains in close proximity to the concavity in the outer wall of the optic cup. The lens cells have elongated greatly in comparison with nearby head ectoderm cells (prospective cornea or epidermis), which remain rather flat at this stage. (Such cells are not seen clearly in this photograph.) The morphogenesis of lens that is seen here is believed to occur because of an "instructive" tissue interaction. What is not clear is whether this early lens has sunk inward because of the hypothesis outlined in Figure 7.10, because of action by microfilaments or microtubules, or because of shifting patterns of cell adhesion. (Courtesy of K.T. Tosney.)

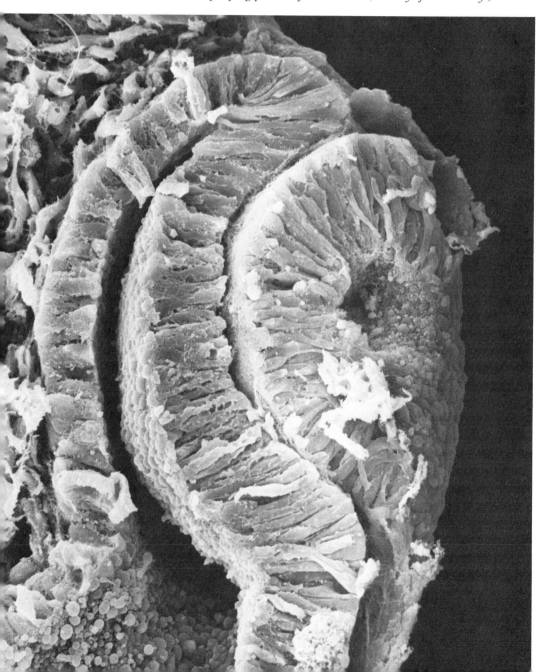

they become firmly attached to one another and no lateral cell movement can occur. If mitosis and a net increase in volume (mass) of that
population of cells occurs, and if the outer surface cannot expand
laterally, then the sheet of cells must buckle inward (as occurs in the
lens vesicle).

Such a phenomenon might account for many cases of folding in
epithelial sheets. The important features are a limitation in lateral
expansion and cell growth. Two obvious candidates for the source of
the limitation can be considered. First, the extracellular materials
known to be present between interacting epithelial and mesenchymal
tissues could act as the attachment site for the cells (i.e., the basal
lamina). Alternatively, the bundles of microfilaments located just
internal to the outer surface of epithelial cells could act as inelastic
bands and could produce the same effect. In fact, bundles of microfilaments oriented appropriately for such a function have been seen
in developing lens, thyroid, lung, and pancreas. Though this scheme,
as proposed by Zwann, is hypothetical, the juxtaposition of "forces"
generated by cell growth and division as they act against a fixed surface (basal lamina?, microfilament bands?) is an attractively simple
mechanism for some types of epithelial morphogenesis.

Besides microtubules, microfilaments, and perhaps specialized
regions of extracellular materials, the adhesivity of cell surfaces could
play a role in the kinds of morphogenesis we have been discussing.
Imagine, for instance, that the composition or quantity of cell surface
materials changes in such a way that the lateral surfaces of epithelial
cells become more adhesive. If you consider the cell shapes seen in
Figure 7.12, this increased adhesivity could cause a greater proportion
of the total cell surface to be allocated to lateral regions, a lesser proportion to end regions. Elongation of the cell would result.

The molecules responsible for intercellular adhesion are under
intense scrutiny. Once they have been identified unequivocally, it
will be necessary to investigate whether their distribution can be regulated so that changes in cell shape can be brought about.

The important point to be understood here is not whether the
hypotheses about microtubules, microfilaments, or adhesivity are
right or wrong, but that we are coming into a position where we can
ask testable questions about the expressive events of early morphogenesis. What causes the organelles to appear at specific times? At
specific sites? In some cells, but not others? What agents affect organelle function, and where do those agents come from? Such questions are speculative. Yet the student should see that they represent

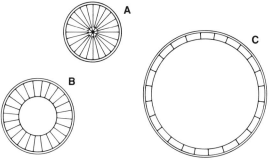

Figure 7.12
Hypothetical effects of changing adhesion between lateral surfaces of epithelial cells. Here three tubules are cut in cross section. The total cellular volume in the three epithelial populations is the same. However, the diameters of the tubules and of the cavities within them are very different, because different proportions of the surface area of each cell are devoted to lateral *as opposed to* end *surfaces. Theoretically, differing adhesions could account for these arrangements of cells: (A) maximal lateral adhesion; (B) intermediate lateral adhesion; and (C) minimal lateral adhesion. Subtle effects arising from the same principle could contribute to the kinds of morphogenesis seen in Figures 7.4 and 7.6 of this chapter. (From F. Gustafson and L. Wolpert,* Biol Rev., 42, (1967), 442.)*

a strategy in which the responding system is dissected and analyzed in detail. Once its functioning is fully understood, then we will be in a good position to predict what classes of control factors should operate. Such control factors must be related in some way to the tissue interactions that are responsible for morphogenesis. The opposite strategy, of course, has been pursued for many years, but with questionable success: go to the "inducing" tissue, analyze what factors it produces, and then search for effects of those factors on responding tissues. Many researchers now suspect that the former strategy may be more fruitful in finally throwing light into the "black box" between interacting tissues.

CONCEPTS

Mesoderm acts permissively in several ways during morphogenesis of branching epithelial organs.

Mesoderm may act instructively to alter epithelial morphology in a very few cases.

Morphogenesis of epithelial sheets involves changes in shape of individual cells.

Localized mitotic control may contribute to epithelial morphogenesis.

Cell elongation, related in some way to microtubules, may con- 103 tribute to epithelial morphogenesis.

Cell narrowing, perhaps caused by microfilament function, may contribute to epithelial morphogenesis.

Changes in intercellular adhesion could contribute to change in cell shape during morphogenesis.

REFERENCES

General:
K. Kratochwil. 1972. In D. Tarin, ed., *Inductive Tissue Interactions and Carcinogenesis.* Academic Press. This is a broad-ranging critique of many different tissue interactions. It argues persuasively for permissive mechanisms; included is literature on lung, salivary gland, mammary gland, and skin.

Nonspecific interaction in the salivary:
K.A. Lawson. 1974. *J. Embryol. Exptl. Morphol., 32,* 469. A classic in careful experimental design that effectively demolishes long-held views on salivary development.

Cell shape control:
American Zool., vol. 13 (1974), no. 4. This issue is devoted to the many ways that cell shape is controlled, and includes papers on microtubules, microfilaments, GAG, collagen, hyaluronidase, and many tissue systems (Bernfield's work on salivary glands is included). A prime source for references.

Nonshifting neighbors in medullary plate:
B. Burnside and A. Jacobson. 1968. *Develop. Biol., 18,* 537. A unique paper that should be a model for further work.

Microfilaments as rigid bands:
J. Zwann and R.W. Hendrix. 1973. *Amer. Zool., 13,* 1039. Zwann's proposal that microfilaments may be inelastic stabilizing agents during morphogenesis.

Lung morphogenesis:
N.K. Wessells. 1970. *J. Exptl. Zool., 175,* 455. Comparison of bronchial and tracheal mesoderm effects, and references to Alescio's pioneering studies.

Adhesion and morphogenesis:
A.A. Moscona, ed. 1974. *The Cell Surface in Development.* Wiley. Several papers in this volume treat aspects of the adhesion problem and how it may affect morphogenesis in model systems and in the embryo.

Chapter Eight:

The cytoplasm of a differentiated mouse embryonic pancreas exocrine cell. The ribosome-studded membranous sacs are the rough endoplasmic reticulum, the organelle where the exocrine secretory proteins are synthesized. Portions of storage granules (zymogen granules) are seen to the lower right. Within them are stored inactive precursor forms of the digestive enzymes (e.g., trypsinogen, chymotrypsinogen). A small portion of the nucleus is at the top left.

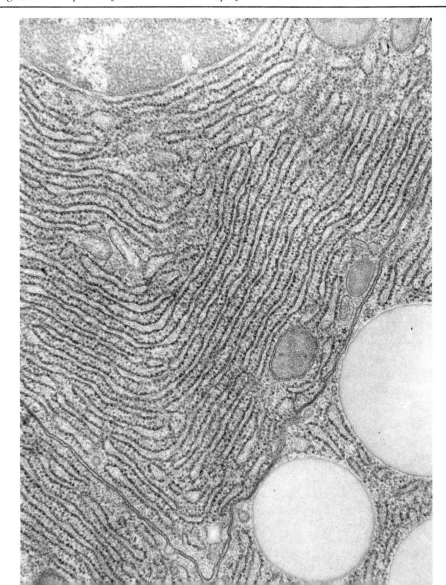

Tissue Interactions and Cell Differentiation

None, then a little, then a lot—the stages in protein-synthesizing activity during differentiation.

In this chapter we will treat development of the mammalian pancreas in detail, since most of the important principles of cell differentiation are well-illustrated by this organ. We will see how morphological and biochemical criteria can be applied in analyzing the differentiative sequence, and how the various events of regulation relate to tissue interactions. Finally, we will consider other examples of endodermal tissue development, and compare the conclusions about tissue interactions with those we reached in Chapter 7 about morphogenesis.

The dorsal pancreas of mammalian and avian embryos arises as a hollow bulge from the upper surface of the gut (Figure 8.1). The inner endodermal layer (an epithelium) grows into the investing mesenchyme and begins to form fingerlike cords of cells, which in turn give rise to spherical clusters of cells called acini. The process of morphogenesis results in a gland that closely resembles a salivary gland in having a complex set of ducts draining secretory acini.

106

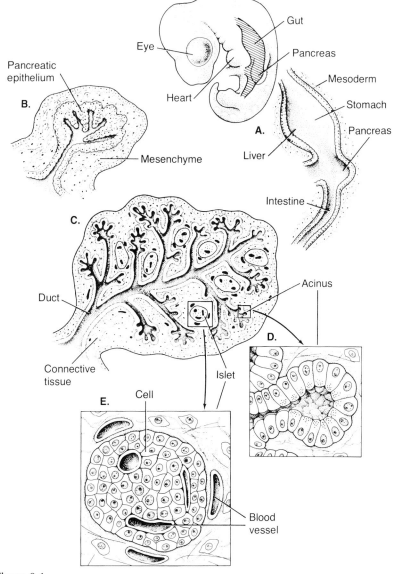

Figure 8.1

*Pancreas development in a rodent. The position of the developing gut and pancreas are shown on the sketch of this 9-day-old mouse embryo. **A.** The initial elevation of the pancreatic epithelium has taken place, and if measurements were made, low levels of specific tertiary proteins (glucagon, lipase, etc.) would be detected. Thus the two expressive events, morphogenesis and cellular differentiation, start together in time. **B.** By about 12 days, the epithelium has grown in size and is beginning to form secretory cell clusters called acini. **C.** By about the fifteenth day of gestation, a section through the main mass of the pancreas might look like this. The abundant acini are connected to a system of tubular collecting ducts. Islets have separated from the epithelium and are in intimate association with blood vessels, into which they can discharge glucagon or insulin. **D** and **E** show enlargements of an acinus, the cells of which contain secretory (zymogen) granules, and an Islet, respectively. (From "Phases in Cell Differentiation" by Bill Rutter and Norman Wessells. Copyright © 1969 by Scientific American, Inc. All rights reserved.)*

A remarkable divergence arises in the fate of the pancreatic endoderm cells. The bulk of them differentiate into exocrine cells that synthesize, store, and secrete the digestive enzymes used in the small intestine. In addition, significant numbers of endodermal cells pinch off from the main population and set up separate masses of cells called Islets of Langerhans. Islet cells differentiate in several distinctive directions, the predominant types being endocrine cells of two kinds: A-cells, which synthesize the hormone glucagon; and B-cells, which make insulin.

We can carry this description into much finer detail. Because of exceedingly sensitive assays for "specific" pancreatic proteins worked out by William J. Rutter and his colleagues, we can detect the appearance and measure the quantity of the secretory enzymes and hormones. Equipped with these tools we can ask two questions: (1) What is the precise time-course of pancreas development? (2) What types of interactions are involved in the development of the organ?

Answering the first descriptive question, Rutter found that midway through development of a rodent embryo, the exocrine- and endocrine-specific proteins are *not* detectable. This may seem like a silly point at first sight, but, in fact, it is of considerable theoretical importance. The same conclusion is reached by using a very different technique: highly radioactive DNA that codes for the mRNA of tertiary proteins (see Chapter 1 for definition; proteins such as hemoglobin, ovalbumin, etc.) is chemically hybridized with the total mRNA fraction from developing cells. If, for instance, hemoglobin mRNA molecules are present in that fraction, they will combine with the radioactive DNA that encodes for hemoglobin, and the combined aggregate can then be identified and measured. The results are that virtually no mRNA molecules that code for a given protein (hemoglobin) are present prior to a certain time in development (see Figure 8.2). These results are significant because they imply that the genome is not "leaky"; the genes for tertiary proteins can be truly inoperative, and do not initiate transcription until a regulatory event turns them on. Prior to these experiments, the possibility had to be entertained that all genes were turned on a little bit in early development, and that differentiation merely involved modulating rates, not selection of new genes to be transcribed.

The next important conclusion about the pancreas links morphogenesis and differentiation. Just when the initial dorsal bulge of the pancreas becomes recognizable (as in Figure 8.3), the first specific proteins can be measured (glucagon, insulin, lipase, carboxypeptidase

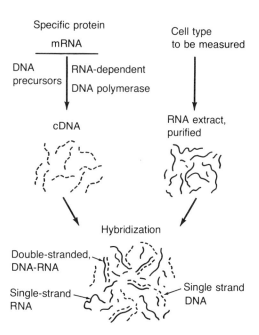

Figure 8.2

Procedures used to measure the quantity of a given mRNA. Highly purified mRNA of a given type is prepared from the polysome fraction of the cell type. Using appropriate precursors and enzyme, DNA complementary (cDNA) to that mRNA is produced (left in diagram). RNA is extracted from the cells to be measured for mRNA content, and, after purification, is chemically hybridized to the pure cDNA. If mRNA that matches the cDNA is present in the reaction mixture, double-stranded nucleic acid is formed which can then be separated and measured (single-stranded RNA and cDNA have physicochemical properties different from those of the double-stranded molecules).

A). Thus, the two expressive processes, morphogenesis and differentiative protein synthesis, appear to be coupled closely in time. Hilfer has seen the same temporal coupling in thyroid development, where morphogenesis starts coincidentally with the first synthesis of the hormone thyroxine.

The following four to five days of pancreas development are characterized by intensive mitotic activity, early phases of morphogenesis, and a fairly constant low level of the pancreatic specific proteins (Figure 8.3). Then nondividing cells begin to accumulate in the exocrine acini. It is within such cells that a large, complex endoplasmic reticulum, an abundant Golgi apparatus, and storage ("zymogen") granules appear. These structural features characterize the "differentiated" secretory cell. In fact, the process whereby these phenotypic characteristics arise is what is meant by "differentiation." By the last few days of gestation in a mouse or rat embryo, virtually all cell division ceases in the pancreas, and most cells appear to be packed with zymogen granules or endocrine granules. After birth, of course, a low level of mitosis is reactivated in the functional gland to permit growth of the organ commensurate with the increase in over-all body size.

Turning to the second question, let us now begin to examine the interactions involved in pancreas maturation. First, if we dissect out the midgut of a mouse embryo about eight hours before pancreas morphogenesis and specific protein synthesis start, and free the endoderm

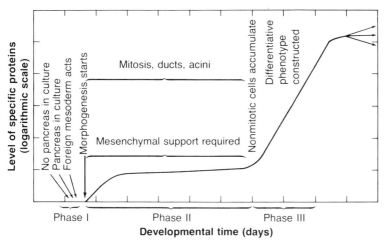

Figure 8.3

A summary of time-relationships in pancreas development. The curve reflects the quantity of proteins specific to exocrine cells that is present during embryonic development. During Phase I, important developmental changes are occurring in the prospective pancreatic population, though at this time no pancreas can be seen. Then morphogenesis and specific synthesis start, and continue through Phase II. Mesenchyme or Rutter's mesenchymal factor is required through most of this period. Next, nonmitotic cells begin to accumulate, and the differentiated phenotype characteristic of the cells in an adult pancreas is constructed.

from adjacent cell types (by using an enzyme bath), we can then recombine the endoderm with foreign mesenchyme (from salivary glands, for example). If the two tissues are placed in organ culture, in the succeeding seven to eight days they can be seen to form a miniature exocrine pancreas, composed of fully differentiated secretory cells arranged in normal acini. High levels of amylase, a specific pancreatic enzyme, are present in the zymogen-packed cells. This experiment permits the conclusion that, *prior* to visible initiation of pancreas development, a discrete area of endoderm tissue requires only permissive support from mesenchyme in order to develop into pancreas.

Does an instructive interaction ever occur in pancreas development? The evidence, some of which is summarized in Figure 8.4, is not clear. What is particularly puzzling is seen in part *a* of that figure, since at the early stage no pancreas forms but liver, lung, stomach, and intestinal-lining tissues all arise from the cultured gut tissues. It is as if no "information" leading to pancreas development is available in the cultured tissue, though information for all the other organ types is used.

In summary, then, the early gut passes through the following stages: (1) a time when cultured pieces form no pancreas; (2) a time

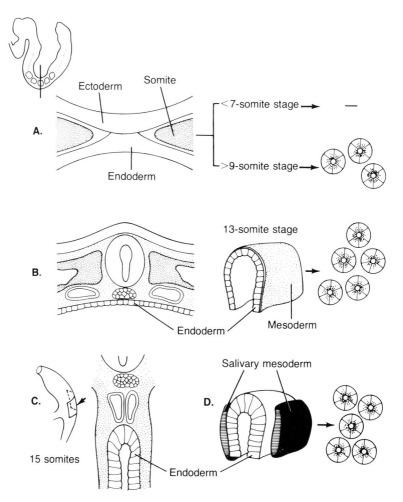

Figure 8.4

*Development of pancreas in culture. A cross section of an embryo with six or less somites (upper left) appears in **A**. If the whole midgut region is cultured from embryos of less than the 7-somite stage, then no acini form. The same tissues from embryos at about the 10-somite stage form acini. By the 13-somite stage (**B**), the notochord and endoderm are separate; if the endoderm and its few adhering mesenchyme cells are cultured, acini form. By 15 somites, no pancreas is yet visible; however, a region of the dorsal gut can be removed (**C**), cleaned of mesenchyme with enzymes, recombined with salivary gland mesenchyme (**D**), and acini will form. Note that all these changes are occurring before the expressive events we can see or measure occur (i.e., morphogenesis, initiation of specific protein synthesis).*

when cultured pieces form pancreas; and (3) a time when endoderm freed of normal surrounding tissues can form pancreas in the presence of foreign mesenchyme. It should be emphasized again that all three stages *precede* any visible indication of pancreas development. We conclude, in particular, that the last stage reflects a condition in which a population of about 300 cells has a marked "stability" for pancreatic development. This stability may mean that the process of restriction is completed, but no successful tests of heritability have been run. Even if this proves to be the case, the source of information leading to pancreas restriction and determination has not been defined.

The Mesenchyme in Pancreas Development

What can we learn about the role of interactions in the remainder of pancreas development, particularly during the period of early morphogenesis and low-level specific protein synthesis? At any time during this period, the pancreatic epithelium can be combined with a variety of embryonic mesenchyme types and exocrine cell differentiation will proceed normally (i.e., the transition from phase II to phase III in Figure 8.3 will occur). In fact, Rutter has found that a cell-free extract of whole chick embryos or of mesenchymal tissues can be added to the nutrient medium in which pancreatic epithelium is cultured, and again differentiation proceeds on schedule. Thus the permissive action of mesenchyme is mimicked by the extract.

How does the extract act? Ronzio and Rutter have discovered that both DNA synthesis and levels of the enzyme DNA polymerase fall drastically in epithelium that is cultured alone. The extract or a protein-containing macromolecule from it causes both enzyme level and DNA synthesis to increase substantially (see Figure 4.2). Pretreatment of the extract with periodate, a substance that oxidizes GAG and proteoglycans (see Chapter 15), destroys the activity. (Periodate would also oxidize the ends of mRNA and tRNA molecules. However, since ribonuclease, the enzyme that degrades RNAs, does not inactivate the factor, it is assumed that the periodate effect is not on RNAs in the factor. This is a kind of compromise that must often be made in experimental biology.) The general proteolytic enzyme trypsin also inactivates the factor. Interestingly, hyaluronidase, collagenase, and neuraminidase, an enzyme that hydrolyzes substances found on the surface of cells, do *not* destroy the active factor. Those enzymes would, of course, remove from the salivary epithelial surface the materials that

112   are required for salivary morphogenesis. These results suggest that the
factor acting on pancreas is not the same as the materials that are essential for salivary gland morphogenesis.

The Mesenchymal Factor and the Cell Surface

The mesenchymal factor (MF) that stimulates epithelial mitosis in
pancreas is believed to act at the cell surface. Rutter and his colleagues
have covalently bound MF to large, insoluble beads composed of a
substance called Sepharose (a derivative of a plant carbohydrate,
agar). When such beads are placed in the presence of pancreatic epithelia, the cells attach to the beads, become oriented (another surface
effect!), and maintain DNA synthesis and cell division (see Figure 8.5).
With further culturing, the cells form zymogen granules, thereby
indicating they have differentiated as exocrine cells.

   Let us consider carefully the evidence that bound MF is active.
First, the pancreatic cells do not attach to beads that lack covalently
bound MF (i.e., beads that are simply soaked in MF solution, under

Figure 8.5

*A section through a mass of pancreatic epithelium attached to Sepharose
beads coated with MF. The dark oval structures are nuclei covered
with silver grains (see legend to Figure 5.6). These cells have
incorporated a radioactive precursor of DNA and are carrying out
cell-division cycles. If similar beads were present but lacking the MF,
then only rare epithelial cells would synthesize DNA or divide.
(Courtesy of R. Pictet and W. J. Rutter. From* Nature New Biology,
246 (1973), 49.)

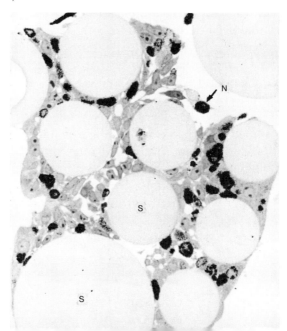

conditions where no covalent binding occurs, are inactive). Second, during the culture period, MF is not released from the beads into the culture medium. Lastly, the pancreatic epithelia show their response to MF on beads when the total MF available is far less than would be needed if soluble MF were simply present in the nutrient medium. In other words, binding of MF to Sepharose effectively concentrates it; so less is needed to stimulate the epithelial cells.

Rutter's group points out that MF is normally of low solubility under the conditions of pH and salt concentration that are physiologically typical. Thus the MF molecules might be in the form of an aggregate or complex in the extracellular spaces surrounding the pancreatic epithelium in an embryo. It should not be surprising, therefore, that MF is active when immobilized on beads. What is important is that it can act when so bound. This implies that it functions *at the cell surface* in promoting mitosis. We will see later in this book that other "growth-promoting" substances also are effective when bound to beads. Action at the cell surface may be a general phenomenon.

When one thinks about mitosis and its control, it is natural to speculate about cyclic nucleotides. In cultured cell lines and in some cancer cells, low levels of cyclic AMP and high levels of cyclic GMP are characteristic of dividing cells (Chapter 13). It is also known, of course, that many proteinaceous hormones exert their effects on cells by controlling levels of cyclic nucleotides (to be discussed in Chapter 13). Does MF act in this manner?

Rutter's group finds that the cyclic nucleotides themselves have no effect, pro or con, on mitotic activity in pancreatic epithelia. However, if one inactivates MF by oxidation with periodate, then its activity can be restored with forms of cyclic AMP that can enter cells. Thus, DNA synthesis and mitosis are stimulated by inactive MF plus dibutyryl-cyclic AMP, but not by either agent alone. This result is just opposite to what would be predicted from the work on cultured adult cell lines (see above and Chapter 13). We shall encounter this problem again in Chapters 12 and 13 where it is pointed out that *high* levels of cyclic AMP are found in regenerating liver and limb tissues at times when mitotic rates are high. The first lesson we should remember, therefore, is that it can be false and misleading to extrapolate to embryos the conclusions reached for cell lines in culture (since in them, high cyclic AMP would *inhibit* mitosis). Regulatory mechanisms operating in intact tissues and organs composed of several cell types may be quite different from those working in "model" systems, such

114   as a cell line in culture. We are left then, with the working hypothesis that MF acts at the surface of pancreatic cells, and may require relatively large amounts of cyclic AMP in order to promote cell division.

The simplest interpretation of the results on the pancreas is that the permissive action of mesenchyme stimulates mitosis in epithelial cells. If mitosis occurs, then differentiation can follow. One of the attractive features of this interpretation is that it may also apply to other developing organs. Thus we can propose the generalization: one of the interactions responsible for the over-all development of most organs is mesenchymal support of epithelial mitotic activity. Additional interactions are, of course, needed so that restriction or the more complex expressive processes can be controlled.

We can next ask, how long must the mesenchyme or extract work for differentiation to occur? To answer this question, the mesenchyme or extract is removed from cultured epithelium after various times of exposure. It turns out that the mesenchyme is required during most of the period of mitosis and low-level specific protein synthesis (phase II in Figure 8.3). Then, about 12 to 18 hours before the transition to high rates of specific protein synthesis and cell differentiation, mesenchyme can be removed and no detrimental effects are seen. By that time the epithelium is, so to speak, over the hump; it can proceed on its own to attain the exocrine cell phenotype. These findings make sense when we realize than an important cellular event occurs at the time of transition between phase II and phase III (Figure 8.3; i.e., the time when the rate of specific protein synthesis accelerates). It is at that time that the first *nonmitotic* population of epithelial cells begins to accumulate. It may be no coincidence, therefore, that mesenchyme is needed only until about 12 hours prior to that event, since a 12-hour period corresponds roughly to the length of a single mitotic cycle in cultured pancreas cells. Thus the hypothesis that mesenchyme serves as a mitotic stimulant allows us to understand why that tissue is needed for only a circumscribed period prior to differentiation.

Regulatory Events in Pancreas Development

The transition point to high rates of differentiative syntheses deserves special emphasis. The equivalent point can be recognized in many other differentiating cell types. The pancreatic transition apparently corresponds to: the time when myoblasts cease cell division and fuse to form multinucleated skeletal muscle cells; the time when neurons

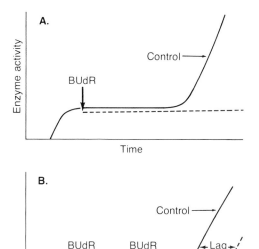

Figure 8.6
*An interpretation of the results of
applying bromodeoxyuridine (BUdR)
to pancreas cultures. In **A**, BUdR
is present continuously, and accumulation of all the exocrine-specific
proteins is inhibited. The effect is
partially reversible, since if BUdR is
present for only about two days (**B**),
then accumulation of the specific proteins
occurs after a lag of about 1.5 days.
In all probability, the cells are essentially
"growing" out of the inhibition in
this case. (Data in Walthar et al.)*

cease cell division, become "specified" (to be discussed later), and send out axons and dendrites; or the time in plasma cells of the immune system when DNA synthesis cycles halt and massive antibody synthesis begins. It is worth our while, therefore, to examine the regulatory events associated with this basic alteration in cellular activity.

The mitotic cycles that immediately precede the transition to high rates of specific protein synthesis are a time of significant developmental events. For instance, if the compound bromodeoxyuridine (BUdR) is added to the nutrient medium during the stage of mitosis (phase II), then differentiation and the acceleration in rate of specific protein synthesis do not occur. The drug is not toxic, since mitotic activity continues for several days. What is so remarkable about the drug's effect is that the synthesis of primary proteins (common cell proteins; see Chapter 1) is not inhibited by BUdR to the same extent that tertiary proteins (secretory enzymes, in this case) are blocked. This differential effect, coupled to the lack of inhibition of RNA, DNA, and lipid syntheses, accounts for the fact that mitosis continues but differentiation is absent.

Wilt and Anderson have reviewed the variety of actions of BUdR in a paper that is a model for those interested in learning how to interpret the results of experiments using drugs. One interpretation of the BUdR work stems from the fact that BUdR is a structural analog of thymidine, one of the bases in DNA. The compound is incorporated into DNA of pancreatic and other cell types. An explanation for

116   BUdR's action may come from work on bacteria, where the compound causes abnormally "tight" binding of the *lac* repressor protein to *lac* operon DNA; an equivalent action in eukaryotic cells might reduce or eliminate the production of mRNAs that code for differentiative proteins.

The importance of RNA synthesis for pancreatic development can be shown by applying actinomycin D, an agent that binds to DNA and inhibits synthesis of various classes of RNA. In the pancreas, the sensitive period to actinomycin is the same as that for BUdR: if actinomycin is applied prior to the transition to high rates of specific protein synthesis, then the acceleration never occurs; if it is applied after the transition, the rate of synthesis accelerates, and the differentiative phenotype is attained. It seems likely, therefore, that if final differentiation is to take place, syntheses of normal RNAs must go on during the period of permissive interaction with mesenchyme.

Some biologists have proposed that unique and essential events occur during the terminal cell divisions prior to such differentiation. If inhibitors of DNA synthesis, and indirectly of mitosis (such as fluordeoxyuridine, FUdR, a drug that prevents thymidine synthesis), are applied to pancreas tissue during phase II, then the transition in synthetic rate and phenotypic differentiation do not occur. Applied later, there is no effect, just as one would expect if there were no side-effects of the drug.

These observations agree with those on other embryonic systems. A speculative working hypothesis is: DNA synthesis is a prerequisite to the transition in rate of specific protein synthesis. We do not know whether this means that some unique kind of DNA is made, whether certain changes in regulatory state can occur only when chromatin is replicated, or whether some other unknown event goes on. Furthermore, the experiments do not *require* us to conclude that there is anything qualitatively different about the last DNA synthesis cycles as opposed to earlier ones. Perhaps the safest conclusion at present is that the normal cell division cycles prior to the regulatory transition must be completed under conditions where normal DNA synthesis can occur, if the transition is to take place.

Finally, it is worth emphasizing a point made when we discussed the stability of the differentiated condition, a point that applies to both pancreas and other organs: once the regulatory transition to a high rate of specific protein synthesis is passed, there is no intrinsic antagonism between mitosis and specific protein synthesis or the differentiated state. Thus mitosis can occur in cells with secretory granules or analogous indications of differentiation.

Let us now summarize some of the arguments about pancreas development. The precise time and cause of determination of the prospective pancreatic cell population is undefined. Morphogenesis and low-level specific protein synthesis begin together, at what we might call the initiation of the expressive phase of pancreas development. A period of mitosis and morphogenesis follows. Finally, transition to increased specific protein synthesis occurs and the final cellular phenotype is constructed. The apparent separation in time between determination and the DNA- and RNA-dependent events that occur days later emphasizes that not all significant regulatory events associated with nucleic acids occur during the restrictive phase of cell development; some also occur during the expressive phase. The ability of nonspecific foreign mesenchyme to support pancreas development before the expressive phase even starts suggests that we need not implicate any foreign "instructional" molecules as causative agents for expressive development of pancreas.

What Triggers the Transition?

What tips the balance and leads to the commencement of phenotypic differentiation? The mitotic population of the expressive phase apparently does not operate on an unalterable time schedule. Normally, either in an embryo or in a variety of culture conditions, pancreatic exocrine cells do follow a timetable. Thus, if a piece of tiny pancreas from an 11-day mouse embryo is removed to a culture dish where appropriate nutrients are supplied, one can predict with great certainty that five days later exocrine cells will be filled with zymogen granules, an extensive rough endoplasmic reticulum will be present, and the rate of specific protein synthesis will be high. The attainment of these features mirrors precisely similar events in intact embryos.

This normal behavior is not obligatory. For instance, if the growing pieces of pancreatic epithelium are subdivided periodically during the culture period (say, at 48-hour intervals), cells remain in a mitotic state and fail to undergo transition to phenotypic differentiation (see Figure 8.7). Kept in this condition for even 11 days (i.e., until a time equivalent to three days after birth of a mouse embryo), cells still retain the capacity to differentiate if they are not disturbed by further tissue subdivisions. These experiments and analogous ones on developing muscle emphasize that the normal timetable of the expressive phase can

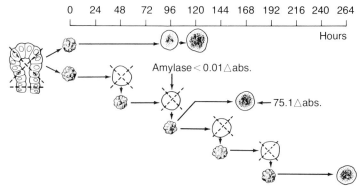

Figure 8.7
*Delay in differentiation due to fractionation of pancreatic epithelium.
Normally, zymogen granules and high levels of exocrine secretory enzymes
(amylase data shown) appear on the fourth and fifth days of culture.
Repeated fractionation of the growing epithelia delays these manifestations
of differentiation. (Enzyme activities are measured in terms of change in
absorbance units in time.)*

be altered. This should not come as a surprise. Recall our discussion of
the *lack* of stability of the phenotype of the differentiated cell. Alterations in the environment can cause dedifferentiation and rapid mitotic
activity. The experiments on subdivision of pancreatic epithelia may
be thought of as a case in which a developing cell population is maintained in an analogous state of mitosis and incipient phenotypic differentiation.

What is it, then, that tips the balance and leads to nondivision and
phenotypic differentiation? In the subdivision experiments, the mass
of the tissue apparently influences the behavior of the component
cells. Reduction of tissue mass near the time of transition delays the
transition. A variety of other kinds of experiments on invertebrate
and vertebrate embryos have suggested the same thing; in fact, differentiation may not occur at all in some systems if too few cells of a given
type are present.

This line of reasoning on the pancreas and other tissues leads to
the idea that the normal increase in mass may cause the transition.
Does this mean that increasing the mass of primitive, 11-day pancreatic
epithelium will elicit precocious differentiation? Alas, life in an embryo is not as simple as our hypothesis. Masses of 11-day epithelia,
larger even than a single epithelium grown in culture for six days,
develop on their *normal* schedule—differentiation cannot be speeded
up by this device (see Figure 8.8)! Again, on afterthought, this should
not be surprising. We know already that DNA- and RNA-dependent

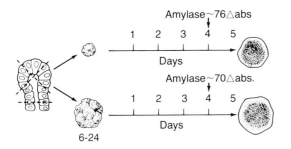

Figure 8.8
Lack of effect of increasing epithelial mass. Clusters of from 6 to 24 pieces of pancreatic epithelium form zymogen granules and amylase activity on the same time schedule as much smaller pieces. In fact, these cultured tissues seem to follow the same schedule that is carried out in the embryo. (Day 4 here corresponds to the fifteenth day of development of a mouse embryo; zymogen granules and amylase would be appearing in intact embryonic pancreas at this time.)

processes must go on during the *expressive* phase if the transition is to take place.

Despite this finding, the mass of a population of developing pancreatic, muscle, or cartilage cells may be a key factor in the transition. We know that such developing cells can alter their own extracellular environments both in terms of the macromolecular complexes present and in terms of the spectrum of physiologically important substances present. Recall the cautionary statements made at the end of Chapter 4 concerning "conditioned" medium and how cells could modify a fluid medium to make it "better" in some sense for supporting developmental processes. It is worth speculating whether the increasing numbers of cells present in developing tissues, either in culture or in an

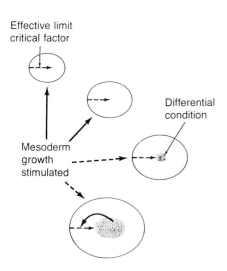

Figure 8.9
Due to the increase in mass of this population of epithelial cells, a differential condition is set up centrally because of the limited range of some factor that is critical for normal development (oxygen? MF? the ZPA substance in limb?). As a result, cells in the central area change in their behavior, perhaps by differentiating. Once such central cells have embarked on a new course, they could act back on the periphery to alter those cells in a variety of ways, including the way that the latter react to the original critical factor. In the real embryo, of course, simple inside-outside relationships, as diagrammed here, are much more complex. Nevertheless, the principle may apply.

120   embryo, do not act in the same way. By altering their immediate environment, they could create new conditions that could serve as a kind of feedback information to cells.

A model system may help to illustrate these points. If one incubates a long-established line of cells in culture (the HeLa strain, for instance) under low population density conditions, then the amino acid serine is required for survival. Serine and a number of other compounds that *can be synthesized* by cells must be added to the medium, because those compounds leak away from the cells faster than they can be synthesized. If we place the same type of cell in culture at high population density, then the cells can meet their own needs, and serine or the other metabolites do not have to be supplied. The reader should be able to extrapolate this situation to a developing mass of tissue in culture, or perhaps even in an embryo. The important relationship is that a large mass of cells can "condition" its own environment in ways that may be impossible for a smaller population of cells. It is not far-fetched at all to imagine that such a conditioning process could include establishment of an intercellular environment conducive to the transition to phenotypic differentiation and to high rates of specific synthesis as seen in the pancreas.

CONCEPTS

The initiation of specific synthesis and of morphogenesis are closely coupled in time in organs such as pancreas and thyroid.

The development of some organs is characterized by stages with differing rates of specific protein synthesis; in pancreas, these are none, low rates, high rates.

Mesenchymes of organs such as pancreas may act permissively by supplying a factor that stimulates epithelial mitosis.

Such a factor may exert its action at the epithelial surface.

Agents that interfere with normal RNA synthesis or cell division cycles inhibit the transition to high rates of specific protein synthesis.

It is not known what triggers the transition to high rates of specific protein synthesis.

Cell division and specific protein synthesis are not mutually exclusive cellular activities.

Under normal conditions, repeated mitotic divisions are not carried out by a differentiated cell; if they are, the differentiated phenotype is lost.

REFERENCES

Pancreas developmental biochemistry:
R.L. Pictet *et al.* 1975. In H.C. Slavkin and R.C. Greulick, eds., *Extracellular Matrix Influences on Gene Expression.* Academic Press. Experiments on cyclic nucleotides, the mitotic factor, and other aspects of pancreas development.

Bromodeoxyuridine and differentiation:
B.T. Walthar *et al.* 1974. *J. Biol. Chem.*, *249*, 1953. This is a model paper in showing how to analyze effects of an inhibitor on a developing system. It includes abundant references on the BUdR effect, the lac operon, and pancreas development. See also F.H. Wilt and M. Anderson. 1972. *Develop. Biol.*, *28*, 443, for an even broader review and critique.

Thyroid morphogenesis and specific synthesis coupling:
W.G. Shain *et al.* 1972. *Develop. Biol.*, *28*, 202. This paper shows the temporal coupling of these two expressive phenomena.

Mass effects and differentiation:
N.K. Wessells and J.H. Cohen. 1967. *Develop. Biol.*, *15*, 237. Effects of fractionation or combination of pancreatic epithelia are reported. (See also discussions in the books by Berrill and Carp, and by Lash and Whittaker, in the references for Chapter 1.)

Hemoglobin and blood cell development:
R.A. Rifkind. 1974. In J. Lash and J.R. Whittaker, eds., *Concepts of Development.* Sinauer. See also various papers in Cold Spring Harbor Symp., *Quant. Biol.*, vol. 38 (1973). Hemoglobin mRNA, genes, and erythropoietin are among the topics discussed here.

Conditioned medium effects:
H. Eagle. 1965. *Science*, *148*, 42. This paper summarizes diverse experiments that elucidate controls for cell metabolism.

Chapter Nine:

A section through a chick embryo looking upward and anteriorly into the gut cavity. The endoderm cells seen here at the top of the forming gut are destined to form the dorsal pancreas and the stomach; those seen toward the bottom of the photograph will form stomach and liver. (For orientation, the spinal cord, notochord, and paired dorsal aortae are seen dorsally.)

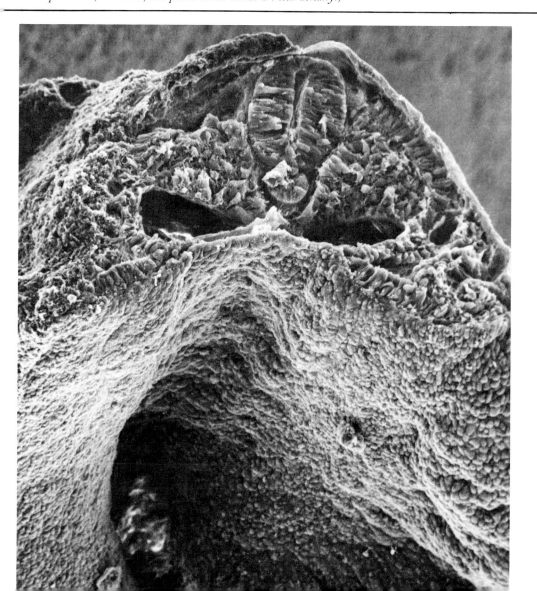

Instructive Properties of Gut Mesoderms

"Your skin is so slippery," he said.
"Yes, my gizzard got displaced!"

The origin of the information that leads to the restriction and determination of various endoderm populations is unknown. In the liver, for example, sequential interactions of endoderm with prospective heart mesoderm and then liver mesoderm appear to be essential. That is, if the endoderm tissue destined to form liver is grafted into contact with prospective liver mesenchyme alone then no liver forms. However, if the same endoderm is grafted with pieces of both precardiac and preliver mesenchymes, then liver epithelial tissue develops. This experimental combination mimics normal liver development in a sense, since the endoderm that will form liver passes through a transient phase of contact with precardiac mesenchyme.

Ideally, one would like to apply all the criteria for instructive tissue interactions listed in Chapter 4 to this case of liver development. So far, heart mesoderm has not been shown to be able to cause any other part of the endoderm layer to form liver (recall the transplanted optic

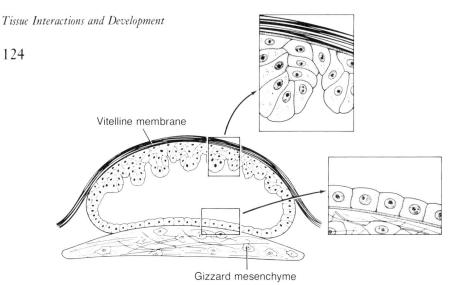

Vitelline membrane

Gizzard mesenchyme

Figure 9.1
Differing responses of an epithelium to adjacent substrata. Here rabbit stomach epithelium is placed in contact with gizzard mesenchyme on one side and a nonliving sheet (the vitelline membrane from a chick egg) on the other. An epithelium that is one cell thick, with regularly shaped and regularly spaced cells, results in the first side. The epithelium in contact with the vitelline membrane undergoes mitosis and forms mounds of cells reminiscent of the surface epithelium elsewhere in the gut. If another type of mesenchyme were present in this culture instead of the vitelline membrane, then still a different pattern of epithelial development might occur. The fact that different patterns of development can be carried out in separate portions of a single epithelium emphasizes the localized nature of the tissue interaction and response. (After D. David, W. Roux Arch. Develop. Biol. 170 *(1972), 1.)*

vesicle causing head ectoderm to form lens). Similarly, for thyroid, pancreas, lung, and portions of the gut tube proper, the full set of experiments that would define instructive tissue interactions has not been carried out. So, for the major gut derivatives—liver, pancreas, lung, etc.—we are left with the conclusions of the preceding two chapters: permissive interactions occur and are essential for development of such organs.

It is true, however, that some of the mesodermal subpopulations associated with different segments of the gut system have instructional capabilities. Suppose we remove mesoderm from the early esophagus, gizzard, intestine, or lung, and combine the tissue with a "neutral" epithelium. One neutral tissue used for such experiments comes from the allantois, a sac-like organ that grows out from the posterior embryonic gut. When allantoic endoderm is combined with various of the gut-associated mesenchymes, it differentiates into tissue that appears very similar to that of the area from which the mesenchyme

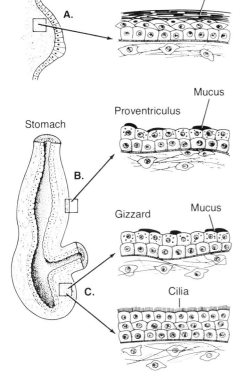

Figure 9.2
Variations in epidermal differentiation as a function of mesenchyme type. Epidermis from the back of a chick embryo is combined with mesenchyme from (A) limb dermis; (B) proventriculus (the anterior stomach), or (C) gizzard (the posterior stomach). After several days of incubation, the epidermis has formed keratin when dermis is present (A), or mucus when the gut mesenchymes are nearby (B, C). In some cases (C), a ciliated outer layer forms, the cells of which may secrete mucus. (Redrawn from McLoughlin.)

was removed. Thus, with gizzard mesenchyme, a thick, "stratified" epithelium forms and contains glycogen granules; with intestinal mesenchyme, simple columnar cells form, just as in the intestine; or, with proventricular mesenchyme, a cylindrical epithelium containing glands develops. In a different kind of recombination, mesenchyme from the gizzard causes *ectoderm* from a young chick embryo to form ciliated, mucous-secreting cells (see Figure 9.2). This is a striking diversion from its prospective fate of forming keratinized (containing "keratin" proteins) epidermis or feathers. Many alterations like these have been caused in the gut endoderm itself, if mesenchymes from different regions are interchanged (intestine to stomach, etc.). Finally, one can even perform these sorts of experiments by using epithelium from an endodermal derivative such as lung. When early lung is combined with various gut mesenchymes, bronchial branching ceases; however, the epithelium thickens to varying degrees and takes on characteristics of the respective gut regions.

If we accept somewhat crude morphological criteria, and rather imprecise biochemical analyses, as indices of differentiation, then we are forced to conclude that instructive interactions are occurring.

However, what does not apparently occur in such recombinations is an effect on complex branching morphology. Thus, a branching

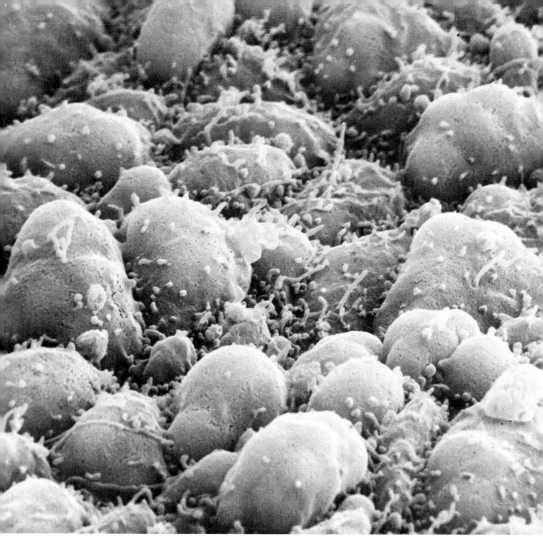

Figure 9.3
Endoderm cells lining the chick embryonic gut in the region where the stomach will form.

lung system is never caused to form from a neutral epithelium or from nonlung portions of the gut endoderm. Similarly, thyroid, liver, or pancreas development cannot be elicited from any but the normal prospective areas of the endoderm system. Either instruction leading to those sorts of restriction and determination does not go on, or all the tissues tested have happened to be incompetent (see Chapter 4) to respond.

We are left with the conclusion that a class of instructive inter-action can be carried out by gut mesenchymes. No tests have been applied to find out how heritable the induced condition in the respond-ing cells may be. Consequently, it is safest at the moment to place these interactions in the same class as the experiment with duck dermis and chick feather, where morphogenesis was under mesenchymal control. Expressive processes (differentiation, morphogenesis) are being effected; whether restriction and determination are also controlled is unknown.

CONCEPTS

Gut mesenchyme can control local characteristics of gut epithelium, such as the arrangement of cells, the presence or absence of glands, or the production of secretory substances.

It is not clear what causes determination of endodermal cell populations of gut and its derivatives (pancreas, liver, thyroid, etc.).

REFERENCES

Liver interactions:
N. Le Douarin. 1964. *Bull. Biol. France, Belg.*, *98*, 589. A monograph-length treatment of tissue recombinations and grafts in developing chick liver. (In French.)
E. Wolff. 1970. *Tissue Interactions during Organogenesis.* Gordon and Breach. A more recent treatment of work on the liver, as well as a summary of the French school's experiments on other endodermal organs.

The allantois as a responding system:
S. Yasugi and T. Mizuno. 1974. *W. Roux Arch. Develop. Biol.*, *174*, 107. Experiments combining a variety of gut-associated mesenchymes with a "neutral" epithelium.

"Self"-differentiation of endoderm:
M. Samiya. 1976. *W. Roux Arch. Develop. Biol.*, *179*, 1. Experiments in which fairly advanced pieces of endoderm are cultured in a nonliving substratum, the vitelline membrane.

Epidermal mucus production in response to gut mesenchyme:
C.B. McLoughlin. 1961. *J. Embryol. Exptl. Morphol.*, *9*, 385. The classic experiments showing ability of gut mesenchyme to act on an ectodermally derived epithelium.
K. Beckingham-Smith. 1973. *Develop. Biol.*, *30*, 263. Observations on the types of proteins made by chick epidermis in culture, and when subjected to conditions fostering mucus production. Potential dangers in extrapolating from culture to *in vivo* situations are documented.

Chapter Ten:

A section through an early tubular gland. Injections of estrogen have elicited the cell division, morphogenesis, and differentiation required to reach the stage seen here. Secretory granules, probably the site of ovalbumin storage, are seen near the apex of the gland cells. (Courtesy of J.T. Wrenn. From Develop. Biol., 26 *(1971), 400.)*

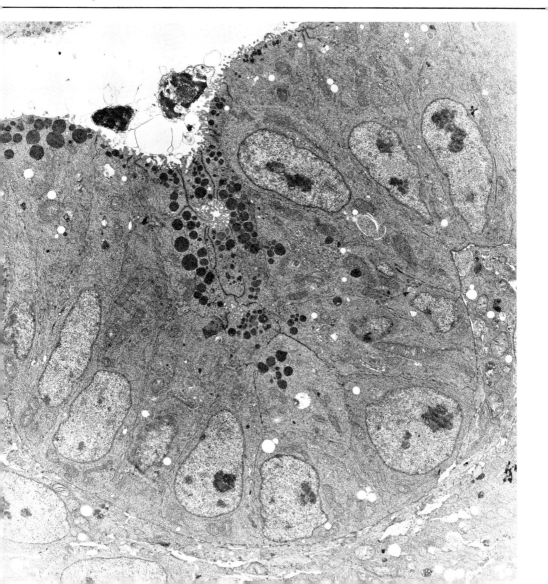

Systemic Tissue Interactions

Hormones circulate everywhere, yet have specific developmental effects. The key is receptor and response.

Not all interactions operate in localized regions of an embryo. Chemicals circulating in the body fluids, the equivalent of hormones in an adult, can have important developmental roles. For instance, the cornea, which, we saw, depended on intraocular pressure for attaining curvature, will become transparent to light only if it is subjected to the action of the hormone thyroxine. The Coulombres showed that early injections of thyroxine caused a precocious dehydration of the cornea, which is a proximate cause of transparency. Alternatively, if an inhibitor of thyroid gland function (2-thiouracil) is injected at the time dehydration should start, that process and transparency are delayed (see Figure 10.2). In a sense, the thyroid and the cornea are interacting.

It is difficult to fit thyroxine into the permissive, instructive dichotomy we have used for other interactions. The hormone is liberated into the body fluids, is presumably distributed everywhere, and has a variety of effects on different tissues. Specificity clearly must reside

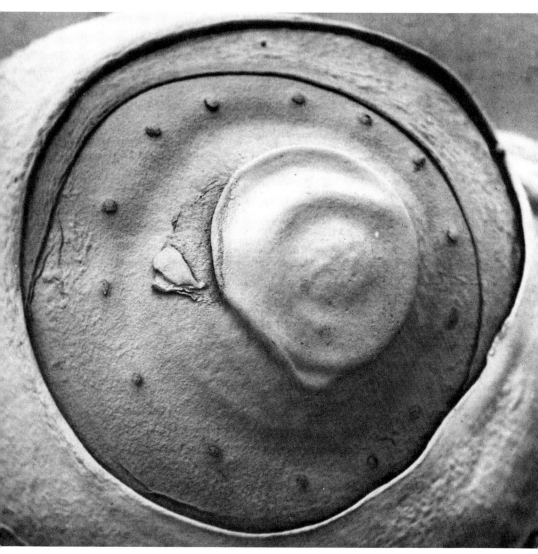

Figure 10.1

The outer surface of the chick embryonic eye. The cornea, located in the center, responds to thyroxine by becoming transparent. The 14 button-like mounds encircling the cornea are the scleral papillae, the aggregations of epidermal cells which act on underlying mesoderm to cause formation of the 14 scleral ossicle bones (see also Figures 4.4 and the frontispiece to Chapter 18).

in responding cells, since at the time corneal cells are responding to the hormone, other cells are not. We can predict, therefore, that one of the differentiative processes in the cornea involves putting cells in a condition where they can respond to thyroxine by carrying out dehydration (a physiological process involving sodium ion transport out of the cells). Thyroxine, then, is a trigger that sets off a vital maturational process. From the organism's point of view, it is as important

130

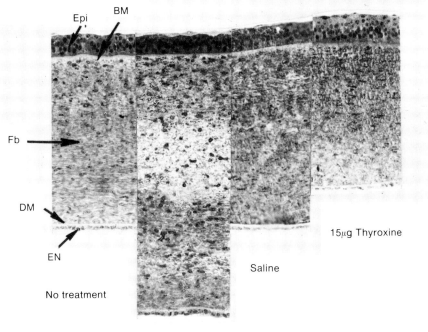

Figure 10.2

A comparison of the thickness of corneas subjected to varying treatments. The nontreated and saline control corneas are alike. Thyroxine has caused a thinning due to dehydration, just as would occur on the embryo as the cornea becomes transparent. Thiouracil, the thyroxine antagonist, causes substantial increase in corneal thickness which correlates with an elevated water content. (Courtesy of E. Masterson, H.F. Edelhauser, and D.L. Van Horn. From Develop. Biol., 43 *(1975), 233.)*

a control factor as the optic vesicle action on lens, or that of lens on cornea.

One might argue, however, that thyroxine is only regulating a common physiological activity, not a unique developmental process. That reservation cannot be raised for a number of other situations where hormones act in development. The steroid hormone estrogen, normally a product of certain cells in the ovary, can elicit both morphogenesis and cell differentiation of oviduct cells. The oviduct of birds is a complex organ responsible for manufacturing the egg "white" and various protective covers for a bird's egg and embryo. One of the main components of the white is a protein called ovalbumin, a substance made in specialized cell clusters in the wall of the oviduct. These "tubular glands" arise in response to estrogen (see Figure 10.3). Even in a little female chick, a few days after hatching, oviduct cells can respond to injections of estrogen by forming tubular glands and by initiating synthesis of ovalbumin.

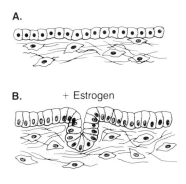

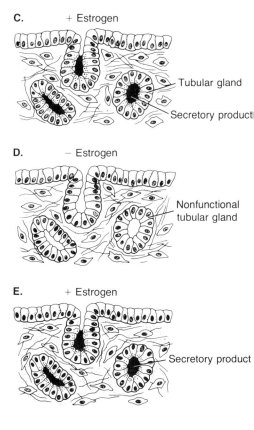

Figure 10.3

A summary of chicken oviduct response to estrogenic hormone. The simple epithelium (A) gives rise to tubular glands (B, C), the cells of which acquire storage granules. Withdrawal of estrogen (D) results in loss of secretory granules and cessation of ovalbumin synthesis. Renewal of estrogen treatment causes rapid resumption of ovalbumin synthesis and reappearance of typical functional morphology. (Redrawn from Schimke).

The morphogenesis of tubular glands seems to include the same sorts of events we described for salivary gland morphogenesis—mitosis, changes in cell shape, microtubules, and microfilaments all may be involved. Similarly, the cellular differentiation process seems to follow the same rules we described for the pancreas. Blocking mitosis (i.e., probably DNA synthesis) with the drug hydroxyurea prevents the stimulation of ovalbumin synthesis by estrogen.

Schimke and his colleagues (1973) have made the important observation that the rate of ovalbumin synthesis during response to the hormone is a direct reflection of the quantity of ovalbumin mRNA present (see Figure 10.5). Estrogen-stimulated cells synthesize the mRNA coding for ovalbumin. The total quantity of such mRNA approaches 78,000 molecules per tubular gland cell, just enough to account for the measured high rates of ovalbumin synthesis.

The manner in which relatively large quantities of a given mRNA accumulate in differentiating cells is complex. Obviously, transcription itself is one means of control. The rate of degradation or turnover may also be important. There is debate about the so-called "half-life"

Figure 10.4

*Three stages in the development of microfilament bundles in chick oviduct cells in response to estrogen treatment. All the pictures are of the outer (lumenal side) ends of the epithelial cells: **A**. 0 hours: no microfilament bundles are present. Ribosomes and other organelles approach close to the inner side of the plasma membrane. **B**. 12 hours after estrogen injection: Microfilaments (M) have begun to accumulate just beneath the plasma membrane. Ribosomes and other organelles are excluded from that region. **C**. 36 hours after estrogen injection: A thick bundle of microfilaments (M) extends across the end of the cell. Probably individual microfilaments insert in the dark-appearing materials at the lateral surfaces of the cell. If so, contraction of the bundle could exert force on the sides of the cell to narrow it at this end and so contribute to morphogenesis of a tubular gland. (Courtesy of J.T. Wrenn. From* Develop. Biol., 26 (1971), 400.)

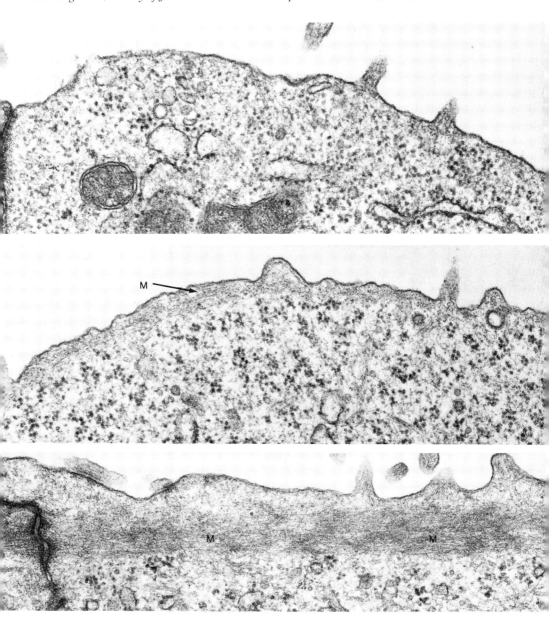

Figure 10.5

Ovalbumin synthesis and estrogen. Initial application of the hormone is followed by a lag as morphogenesis gets underway and tubular glands begin to form. The ovalbumin curve peaks when about 50 per cent of the total protein that is being synthesized by oviduct cells is albumin. If estrogen is withdrawn at that time (10 days on the graph), there is a rapid fall in ovalbumin synthesis. The glands remain in a quiescent state until stimulated once again with estrogen (or another steroid hormone, progesterone). During this "secondary" stimulation, mitosis does not have to occur, and little lag can be detected prior to the start of ovalbumin synthesis. (Redrawn from Schimke.)

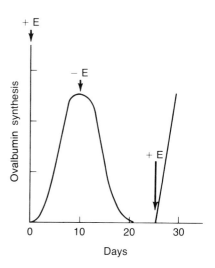

(the time it takes for 50% of an original population of molecules to be degraded) of cellular mRNAs. Some work suggests that mRNA's coding for the common primary proteins may turn over in short times (ca. 3 hrs). However, other techniques suggest that it may be safer to assume that most mRNA of cells from higher organisms have half-lives of 24–48 hrs. Even this is in contrast to mRNA coding for specific (tertiary) proteins; those mRNA molecules persist for much longer periods (as, 100 hours for the cocoonase mRNA of silk moths). This differential stability is not understood but has important consequences, since it can lead to substantial accumulation of a given mRNA even though the rates of synthesis of that molecule are not high. In addition, the phenomenon relates to the common observation that differentiating cells become "insensitive" to actinomycin D, a compound that binds to DNA and interferes with transcription. Recall that after the regulatory transition in pancreas, actinomycin had little effect on the accelerated rate of protein synthesis. The benefit of this "long-lived" mRNA is also seen in cases where cells become functionally enucleated—thus, mammalian erythrocytes can continue to synthesize hemoglobin (because of the stability of the globin mRNA) after their nuclei have been extruded. Besides this peculiar stability of certain mRNAs, some differentiating cells may be able to store mRNA in inactive form, as occurs in most oocytes. These various examples emphasize the complexity of events that may underlie the cause-and-effect relationship between estrogen and increased quantities of ovalbumin mRNA. We will return to this relationship shortly.

The availability of quite pure preparations of individual mRNA types has allowed measurement of the number of gene copies that code for the mRNAs. Chemical hybridization of single-stranded radioactive "cDNA"—the DNA complementary to a given mRNA—with total cell DNA reveals that only one copy (or at most a very few copies) of the genes coding for tertiary proteins (ovalbumin, globin, silk fibroin, etc.) is present. Calculations show that single copies per haploid genome would be sufficient to account for the measured accumulation of mRNA. For these reasons, we can conclude that gene amplification is not an integral part of cell differentiation.

The results on oviduct development can be interpreted as follows. The oviduct undergoes initial development in the embryo, in terms of both *determination* and initial *expressive* events (an important expressive event would be acquiring the capacity to respond to estrogen, that is, to synthesize estrogen receptor protein). Only later in life, with rising estrogen levels, does final maturation occur, resulting in differentiation and morphogenesis.

The capacity to work with a known compound adds immense power to our analysis. For instance, when Schimke halts estrogen injections, there is a rapid fall in ovalbumin mRNA levels and a concomitant decrease in ovalbumin synthesis, and the differentiated tubular gland cells enter an inactive phase. One of the mysterious processes that goes on during this regression is a decrease in the half-life of the mRNA that codes for ovalbumin, so that those molecules turn over at the same rate as mRNA coding for general cell protein. Clearly, estrogen must be present *continuously* for the tubular gland cells to remain phenotypically differentiated and to synthesize ovalbumin at a high rate.

Still another interesting phenomenon is seen if a new regime of hormone injections is started in the same chick after three weeks: large-scale ovalbumin synthesis resumes. And in this "secondary" stimulation, new DNA synthesis is *not* required for the regulatory transition. Here we are seeing a modulation in synthetic activity in a differentiated cell type. Precisely the same modulation is seen in mature exocrine pancreas cells, where variations in diet alter production of the various digestive enzymes. This fact, that the rates at which protein synthesis is modulated in adult cells are basically independent of DNA synthesis, is in sharp contrast to the *dependence* of the regulatory transitions on DNA synthesis that we saw for differentiating cells. Apparently, a different mechanism operates there, when estrogen, or its equivalent in pancreas, muscle, or nerve, triggers the transition.

The mechanisms by which hormones act are under intensive study. In Chapter 13 we will consider the relation of *protein* hormones to cyclic nucleotides (such as cyclic AMP) in cells. Here, we will continue to discuss *steroid* hormones. Hormone "receptor" proteins can be isolated from several adult cell types known to be sensitive to a given

136 steroid. These receptors are small, soluble protein molecules located in the cytoplasm of cells. The steroid-receptor complex can enter the nucleus of a cell and bind to the chromatin. Since typical responses to the hormone may follow this binding, it seems that the experiment may be showing us how a steroid acts on a cell.

However, there are many uncertainties. One stems from the observation that a given steroid-receptor complex may bind to as many as 1,000 different sites on the chromosomes of a responsive cell. That is far more than would be expected if we assume that the steroid-receptor complex binds to specific genetic sites and initiates RNA transcription.

Another situation that complicates interpretation comes from work on the large "polytene" chromosomes of insects. We will be able to understand this situation after a brief review.

Typically, an insect egg develops into a feeding larval stage (the caterpillar). As the larva grows, it sheds its "skin" periodically during periods called molts. After several molts, the larva may develop into a pupa, a nonfeeding stage in the life cycle that may live within a cocoon for several months. Then metamorphosis takes place, and the butterfly or other insect emerges as an adult.

Two types of compounds coordinate these developmental stages of the life cycle. The steroidal ecdysones are secreted from endocrine organs, the prothoracic glands, when a protein is released from the brain. The ecdysones, or molting hormones, have the general action of stimulating synthetic activities that underlie the molts.

A type of terpene, juvenile hormone, is secreted by regions of the brain called the corpora allata. If large amounts of juvenile hormone are present when ecdysone is released, then the larva molts into a larva. If only small amounts of juvenile hormone are present, then the larva molts into a pupa. If *no* juvenile hormone is present, then ecdysone triggers metamorphosis into the adult form. It seems, therefore, that the amounts of juvenile hormone control the *kinds* of synthetic activity (i.e., which genes are used, larval, pupal, or adult) that are carried out in response to the general stimulant, ecdysone.

The other information we need to know before we discuss modes of hormone action concerns chromosome structure. Many types of insect cells contain "polytene" chromosomes. These are chromosomes in which the genetic material has been replicated a large number of times, and in which the individual strands of chromatin are precisely aligned, so that all copies of a given gene (or, more properly, genetic locus) are next to each other. This arrangement results in "banding" patterns of polytene chromosomes as seen after staining or other pre-

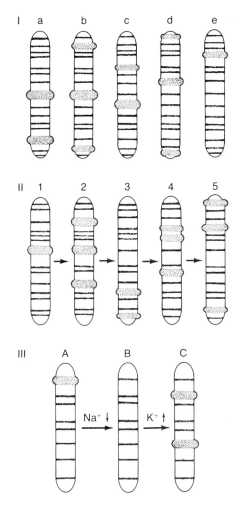

Figure 10.6
Developmental variations in polytene chromosomes of Drosophila. *In I, the same chromosome removed from cells in five different tissues (a, b, c, d, e) is shown diagrammatically. The thin black lines represent genetic loci in the contracted, nonfunctional condition. The heavier black bands represent "puffs," active sites of RNA synthesis. Note that the puffing pattern varies between the five tissues (a-e), reflecting differential use of genes. In II, the same chromosome from a single tissue is shown at five different times in development (1-5). Note that the puffing pattern changes in time, again reflecting variations in gene usage. Finally, in III, we see a single chromosome prior to the addition of ecdysone in A. Ecdysone is then administered, Na^+ levels fall, and some puffs retract (B). Then K^+ levels rise and other puffs appear (C).*

parative procedures. For our purposes, each band may be thought of as a gene.

Normally, the bands of polytene chromosomes are highly condensed structures from which little or no RNA is transcribed. A given band may be activated so that it expands into a "puff"-like structure. The chromatin strands of the band have spread apart to yield the puff appearance. And, at such a site, intensive RNA synthesis can be measured, most likely of precursor molecules of the mRNA encoded in the DNA of the gene at that site.

Suppose we now examine puffing *patterns;* that is, the distribution of puffs in the polytene chromosomes of different cell types or in one cell type at different times in development. We go to a larva, remove various tissues, break open the cells and nuclei, and stain the chromosomes. In tissue *a*, bands "8" and "13" might be puffed. In tissue *b*,

138   bands "2," "8," and "14" are puffed; and so on. Therefore, in these tissues, all sampled *at one time*, there is a tissue-specific banding pattern. This implies tissue-specific gene usage.

Now, let us select a single tissue and follow its puffing pattern through the larval and pupal developmental stages. We see some puffs continuously present, others appearing and disappearing, and still other bands that never puff in this tissue. Again we are seeing evidence of changing gene activity in time. We can conclude that the pattern of gene activity reflected in puffing varies (1) between tissues and (2) within a tissue over time.

If we now perform the experiment of injecting ecdysone into a larva that has a small amount of juvenile hormone present (see above), or if we treat a gland from such a creature in culture with the ecdysone, then the following result is seen. The puffing pattern changes from the larval to the pupal type. An early response to the steroid is appearance of certain puffs. If we follow the hormone-treated tissues over time, other puffs will gradually appear, the early ones regress, and so on. A sequence of puffing activity like that seen in normal development is initiated by the ecdysone. Experiments with drugs that inhibit RNA or protein syntheses imply that RNA is made on the "early" puffs, and translated into proteins, which in turn activate the "later" puffs. Thus, the ecdysone may act as an initiator of sequential gene activity and may *not* have to act on each of the genetic sites that ultimately becomes active in RNA synthesis.

What is the mechanism of action of the steroid in this lovely case? Some workers suggest that hormone-receptor protein complexes may be involved. Others provide evidence that an early response to ecdysone is an increase of amounts of cytoplasmic and nuclear potassium ion (K^+). If instead of injecting ecdysone, we increase the amount of K^+ artificially in cells or in *isolated* nuclei, then the puffing sequence characteristic of ecdysone is started. The hormone itself does not have to be present to elicit the gene activity; K^+ can do it! Other work shows that some early larval bands only remain puffed when sodium ion levels are high; treatment with ecdysone lowers sodium, and those early puffs regress (and, of course, the others responsive to rising K^+ puff out).

These results suggest that one mode of steroid hormone action is on membrane permeability to metal ions. Recall, however, that ecdysone is thought of as a general stimulator of synthetic activities incident to molting. It is juvenile hormone that governs whether larval, pupal, or adult genes will be used when ecdysone acts. Unfortunately,

The complexity of the oviduct's response to estrogen

Stimulus	Responses
Estrogen	mitosis
	microfilament assembly (manufacture?)
	microfilament, microtubule function
	tubular gland morphogenesis
	ovalbumin mRNA production
	ovalbumin synthesis
	ovalbumin mRNA stabilization
	(conalbumin, lysozyme syntheses)

Besides the many responses to the hormone described in the text, the tubular gland cells also synthesize other egg white proteins; thus comments about ovalbumin and its mRNA would apply equally to conalbumin and to lysozyme.

little has been done to establish how juvenile hormone acts on cells. We might guess that it should interact with receptor proteins and with specific bands on the chromosomes—but guesses do not substitute for results. This experimental system deserves intensive study.

With this discussion of modes of hormone action in insects as background, let us now return to estrogen and the oviduct. Some workers have been tempted to conclude that estrogen only acts directly on the genetic machinery in eliciting developmental responses. As Schimke and others have cautioned, however, there is still a black box between hormone administration and tissue response. Since morphogenesis, differentiation, quantities of certain mRNAs, turnover times of mRNAs, and no doubt other phenomena all follow estrogen treatment, it is not fair to conclude that the hormone acts solely at the level of transcription control. Clearly, this is a smaller black box than the ones we have dealt with in other tissue interactions, but it still exists.

The experiments on estrogen and oviduct add some new dimensions to our concept of tissue interactions. A single substance initiates an irreversible process (gland morphogenesis and initial cellular differentiation) and controls a reversible process (synthesis of specific protein and maintenance of the differentiated phenotype). Does the latter, stimulatory role mean that other differentiated cell types are continuously stimulated by regulatory agents? Recall our discussion of the relative *instability* of the differentiative phenotype; perhaps injuries or alterations in local tissue environments are the equivalent of estrogen withdrawal. Dedifferentiation would be analogous to

140   tubular gland regression. If this reasoning is correct, then one of the important components of organ development is the construction of local tissue environments that will stabilize cell phenotype and activity (i.e., the equivalent of a continuous dose of estrogen is acquired!).

On the other hand, it may be incorrect to extrapolate from the oviduct to other systems. The reproductive organs are unique because they are called into play only at a specific time, long after other bodily functions become operational. Furthermore, in most animals reproduction is a cyclical phenomenon that is coupled to environmental variables. One might expect, therefore, that special regulatory mechanisms have evolved to control both the tardy final development and the cyclical functional activity. Only time, and more experimentation, will tell.

Sex hormones can act at the level of tissue interactions. Early mammary gland development is identical in male and female mouse embryos. Then, at a specific time in development, the presence of the male sex hormone testosterone causes the mammary epithelial tissue to degenerate. An experimental demonstration of the hormone-tissue interaction in males and females is outlined in Figure 10.7. In particular, note that testosterone causes female glands to regress in the male fashion—a male "Y" chromosome is not required for that response. Kratochwil has also done tissue recombination experiments involving hormone-insensitive mutant and normal mammary gland tissues (recall the tissue recombinations of polydactylous and eudiplodious tissues in limbs). Kratochwil finds that testosterone only need act on mesenchyme cells. Once activated by testosterone, mesenchyme cells condense around the epithelial stalk of the mammary gland and somehow cause its separation from the epidermis and its subsequent degeneration (i.e., the same stages of regression seen in Fig. 10.7). (The student should attempt to work out on paper how such mutant and normal tissues can be employed to establish this point. Hint: hormone-*insensitive* epithelium combined with male-genotype normal mesenchyme shows mammary degeneration.) This case may be the clearest demonstration known that a specific extrinsic molecule (testosterone) can trigger a typical embryonic tissue interaction leading to a developmental response. We would like to know whether estrogen acts analogously on the oviduct, say by affecting mesenchyme cells so that they stimulate tubular gland morphogenesis and differentiation.

Let us now attempt to relate these hormone-tissue interactions to the sorts of tissue interactions discussed earlier in this book. The common features of all hormone interactions are their systemic delivery

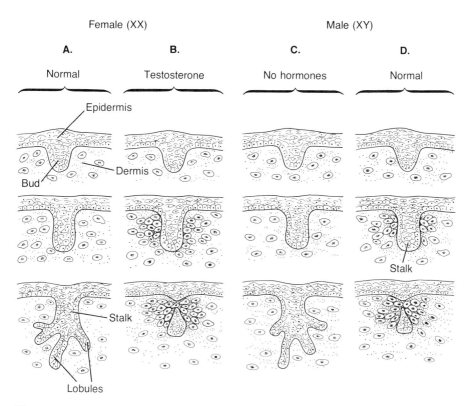

Figure 10.7

*The complexity of interactions between hormones and genetic information is seen in development of early mammary glands. In a female mammal **(A)**, an epithelial bud grows downward into the dermis and branches into lobules. In a normal male **(D)**, the bud forms and elongates, but then mesenchyme cells accumulate on each side of the stalk. The stalk breaks free from the epidermis, and the lower portion of the mammary epithelial tissue regresses and may disappear. If a male gland is cultured in medium lacking hormones **(C)**, then the stalk remains connected to the epidermis and lobules form. This shows that the genetic information required to construct an early mammary gland is present in an individual having the male XY sex-chromosome complement. It also shows that* female hormones are not required *for the gland to develop in the female pattern. Now, if a developing female gland is cultured, either in the presence of testes tissue or with the male hormone testosterone in the medium **(B)**, then regression identical to that in a male occurs. This means that a Y chromosome is not needed in order for the regression phenomenon to occur. It also implies that male hormones are normally responsible for regression of mammary glands during normal development of male individuals. Perhaps the most important thing to emphasize is that the genetic sex of the gland tissue itself has nothing to do with its developmental capabilities. (After K. Kratochwil,* J. Embryol. Exptl. Morphol. 25 *(1971), 141.)*

142 route and the requirement for specificity in the responding cell populations. These interactions seem to contrast with the apparently localized nature of instructive interactions (as in lens or feathers) and certain permissive interactions (as in apical ridge outgrowth effects) in vertebrate embryos. The message "construct leg-type feathers" does not, we assume, circulate everywhere. The orienting effect of the apical ectodermal ridge, at best, must decline precipitously toward the proximal part of a limb. But the reader will recognize our logical difficulties here: it would be possible to assume a localized source (for example, for conditions eliciting "leg" feathers) and systemic distribution, but high thresholds of response except in certain key areas. Only cells in those areas would be subjected to suprathreshold amounts of the critical factor. Apparent "localization" would be the result. We are not just playing games of reasoning. Current techniques and our ignorance of the molecular basis for most interactions simply will not permit hard distinctions to be drawn between localized tissue interaction in an embryo and systemic hormonal interaction.

CONCEPTS

Some hormones play essential and permissive roles during normal tissue development.

A hormone such as estrogen can act both to initiate a developmental process (tubular gland morphogenesis) and to regulate a differentiative function (ovalbumin synthesis).

Steroid hormones of insects may alter the puffing patterns of polytene chromosomes. The mode of action of one such hormone, ecdysone, may be indirect with respect to those chromosomes, since changing levels of sodium and potassium ions may be involved in the response.

Though a given hormone may cause an increase in a specific type of mRNA, the hormone may not necessarily affect transcription of that mRNA directly, nor is its action necessarily limited to effects on nucleic acid synthesis.

Thyroxine and the cornea:
E. Masterson *et al.* 1975. *Develop. Biol.*, *43*, 233. A recent reexamination of the thyroxine and thiouracil effects on cornea; includes measurements on sodium and potassium content, dehydration, and corneal structure.

Ovalbumin and estrogen:
R.T. Schimke *et al.* 1974. In *Recent Progress in Hormone Research.* Academic Press. A recent critical review pointing out the danger of oversimplifying about steroid hormone action.

mRNA and complementary DNA:
G. Brawerman. 1974. *Ann. Rev. Biochem.*, *43*, 621.
D. Sullivan *et al.* 1973. *J. Biol. Chem.*, *248*, 7530.
S. Humphries *et al.* 1976. *Cell*, 7, 267.
These papers summarize knowledge of mRNA, document the methodology and literature on use of cDNA, and give specific data on ovalbumin and hemoglobin mRNAs.

Imaginal discs:
H.A. Schneiderman and P.J. Bryant. 1971. *Nature*, *234*, 187. An excellent introduction to insect developmental biology, with emphasis on imaginal discs and use of mutants.

Insect hormones:
H. Kroeger. 1968. In W. Etkin and L. Gilbert, eds., *Metamorphosis.* Appleton-Century-Crofts.
H. Schneiderman. 1972. In *Insect Juvenile Hormones.* Academic Press. These papers are a good entry to the great amount of literature in this field.

Chapter Eleven:

A single parasympathetic neuron growing in culture is seen in the top photograph. The rounded cell body (at left center) is the site of most nucleic acid and protein syntheses. Building blocks for cytoplasmic organelles are thought to be transported down the long neurites (axons, dendrites) to the active tips (called "growth cones"). The locomotory growth cone is enlarged in the lower picture. The fine cell extensions, termed microspikes, are dynamic structures that move about and seem to explore the environment through which the growth cone moves. The debris that has accumulated on the upper surface of the growth cone may be localized in that region because of cell surface movements that go on as a normal part of locomotory activity by the growth cone.

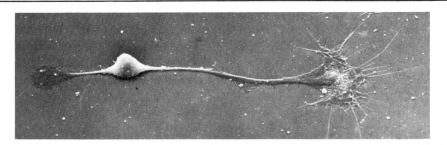

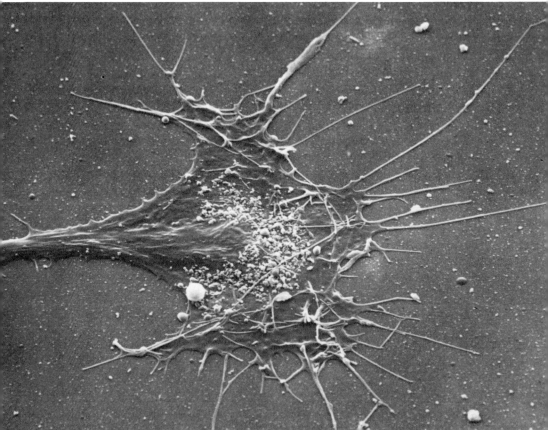

Nerves and Tissue Interactions

That sweet taste of sugar is due to much more than nerve impulses going to the brain.

We have seen that the endocrine system can participate in developmental events. Does the other main integrative system of adult higher animals, the nervous system, act on developing cells? Two types of reaction to nerves will be considered: first, growth in response to nerves; and second, maintenance or generation of the differentiated state in response to nerves.

Regeneration and Nerves

If the distal half of a salamander's limb is removed, regeneration of the missing parts can occur in ensuing weeks. Following initial wound healing, a thickened cap of epidermal cells, reminiscent of the apical ectodermal ridge of the embryonic limb, arises at the point of regenera-

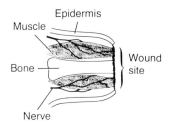

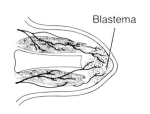

Figure 11.1

Early stages in salamander limb regeneration. After initial wound healing, epidermis covers the amputation site. Then cells that are believed to be derived from the various differentiated cell types near the amputation site (bone, muscle, nerve sheath, etc.) form the blastema. This mass of visibly undifferentiated cells undergoes intense mitosis and gives rise to the regenerated limb.

tive outgrowth (see Figure 11.1). Underlying mesenchymal cells undergo a process of dedifferentiation, engage in intensive mitosis, and yield a "blastema" or mass of mesenchymal cells that ultimately forms the new muscles, bones, tendons, etc.

The regenerative process is dependent on nerves. When the distal limb parts are removed, nerve axons are cut. These axons "regenerate" and grow into the mesenchymal and epidermal areas, where mitosis must produce the cells for limb regeneration. Such nerves probably stimulate that mitosis. If the nerves are removed from one of the forelimbs of a salamander, and then, after the animal has recovered from the first operation, the front feet are removed from *both* forelimbs, regeneration occurs on the leg that has nerves but not on the limb that lacks nerves.

Figure 11.2

An interpretation of Singer's hypotheses about nerve cross-sectional area in relation to limb cross section in three vertebrates. The salamander exceeds the "threshold" ratio and can regenerate its limbs. The lizard is close to the ratio, but does not normally regenerate a limb. Nevertheless, addition of extra nerve to lizard limbs has permitted at least abortive regenerates to form. Mammals have nerves with small diameters and are not close to the theoretical threshold. It would be of obvious medical use if the equivalent of nerve could be added to a mouse or human limb stump so that the ratio could be exceeded. (Redrawn from K. Rzehak and M. Singer, J. Exptl. Zool., 162 (1966), 15.)

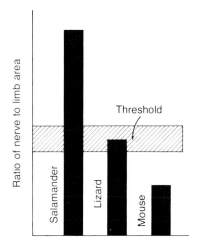

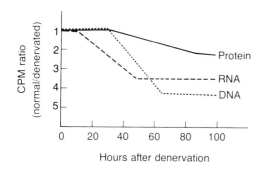

Figure 11.3

Decreased syntheses following nerve removal. Nerves were removed from limbs that were well underway in the regeneration process. At various times thereafter, radioactive precursors of DNA, RNA, or protein were injected into the animals. Four or five hours later, the denervated regenerating limbs were compared with innervated regenerates to see how much of the precursors had been incorporated into the three classes of macromolecules. The measurements are expressed as a ratio of radioactivity (counts per minute, CPM) in normal versus regenerating limbs. It is seen that synthesis of all three declines in the hours following denervation. In other experiments on such animals, it has been shown that removal of nerve from normal, nonregenerating tissues has no effect on these sorts of syntheses. (Redrawn from M.H. Dresden, Develop. Biol., 19 (1969), 311.)

Are specific kinds of nerves required for limb regeneration? Surprisingly, sensory, motor, and autonomic system nerves all support that process. Thus the effect is a nonspecific one. There is a quantitative factor involved, however. Singer and his colleagues have shown that regeneration occurs only if the ratio

$$\frac{\text{total cross-sectional area of nerves}}{\text{cross-sectional area of limb}}$$

exceeds a certain value. We assume this means that a given *volume* or *mass* of nerve axonal material is the crucial factor in regeneration.

Nerves may act in several distinct ways during regeneration: by supplying "trophic" factors and by supplying sheath cells to the blastema. The trophic effect is on various synthetic processes. If we remove nerves from a normal limb, there is no effect on RNA or DNA synthesis. In contrast, when nerves are dissected from an actively regenerating limb, there is a precipitous decrease in synthesis of both RNA and DNA within 40 hours of nerve removal (see Figure 11.3). Protein synthesis is also stimulated in regenerating limb cells by the presence of nerves.

Besides these general effects on syntheses, nerves may act by affecting production and degradation of the important extracellular substances called glycosaminoglycans (GAGs: see Chapter 15 for description). During the early phases following amputation, extracellular GAG, mainly hyaluronate (see Chapter 15), accumulates.

148   Not unexpectedly, this is the time of dedifferentiation, mitosis, and possible cell migration as the blastema forms. Later, hyaluronidase (an enzyme that digests hyaluronate) appears, hyaluronate is degraded, and synthesis of GAG shifts to chondroitins, the major component of cartilage matrix. These changes correlate with cellular differentiation and organ formation (bones, muscles, etc.). We shall encounter the same relationship in the developing cornea (Chapter 15), specifically: hyaluronate is present when cells move and divide; then hyaluronidase appears, hyaluronate is removed, and cells differentiate.

Limbs that lack nerves fail to incorporate normal amounts of precursors into hyaluronate, and subsequently develop little hyaluronidase activity or synthesis of chondroitin. If a limb is amputated and the nerves removed *after* the blastema forms, then regeneration goes on quite normally. Hyaluronidase appears on schedule ten days later. Thus, nerves do not have to be present just before enzyme activity rises. These results imply that the action of nerves is complex and has many indirect consequences. It seems safest to conclude, therefore, that the presence of nerves in an amputated limb stump allows a series of syntheses and other events to go on which leads to the formation of the blastema; that cell population has considerable stability and can carry out its own timetable for subsequent regeneration.

Since RNA, DNA, protein, and GAG syntheses in regenerating limbs depend on the mass of nerve, it is reasonable to ask if the actual "trophic" molecules that stimulate those syntheses can be isolated and identified. M. Singer and his colleagues have partially purified a substance from whole brains or from brain nerve endings (the site where neurotransmitters and trophic molecules might be stored). This material, which may be a basic protein, stimulates protein synthesis in denervated limb regenerates. Since equivalent fractions of basic proteins from nonneuronal tissues do not stimulate such protein synthesis, it may be that Singer's protein will prove to be a trophic factor during limb regeneration.

Despite these apparent successes in linking nerves to synthetic events and in tracking down the hypothetical trophic factor, it is still possible that nerves act indirectly on the synthetic processes in a regenerating limb. Smith and Wolpert point out that the early blastema of the amputated limb lacks blood vessels. New vessels only invade the blastema if nerve is present. Once vascularization is well underway in a blastema, then the nerve supply to the limb can be removed and reasonably normal regeneration will go on anyway. One can speculate, therefore, that nerves establish conditions conducive to blood vessel growth and development, and that such vessels supply required oxygen, nutrients, and hormones to the blastema. Synthetic activities, mitoses, and other regeneration phenomena can then take place, and a new arm or leg appears. (This explanation, linking nerve growth and blood vessel development, is worthy of serious consideration in other developmental situations. We know very little about the rules governing blood vessel development or distribution in embryos; however, their potential importance, once present, should not be underestimated in constructing explanatory hypotheses of organ morphogenesis and cellular differentiation.)

Another contribution of nerves to regeneration is seen when limbs are irradiated with x-rays to the point that chromosomes are so damaged that regeneration cannot take place. The capacity to regenerate a limb is regained if lengths of unirradiated nerve are implanted at the site of amputation (see Figure 11.4). The axons of such lengths degenerate, since they are not connected to nerve cell bodies, but the

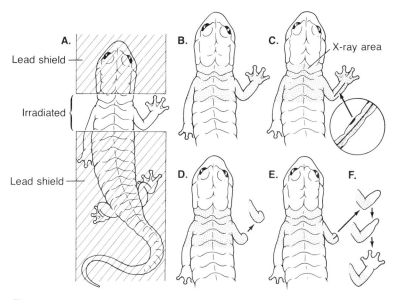

Figure 11.4
*Nerve as a source of cells for regeneration. In **A**, a salamander is shielded with lead except for the forelimb region, which is exposed to sufficient x-irradiation to completely inhibit limb regeneration (note that the nerves and Schwann cells in this region are irradiated too, so that they cannot support regeneration). Next, a length of unirradiated nerve axons and surrounding Schwann sheath is implanted in the irradiated forelimb of one salamander (**C**). Another (**B**) does not receive an implant. The forelimbs are removed from both creatures (**D, E**) with the amputation going through the site of the unirradiated nerve implant. Subsequently (**F**), the limb with the unirradiated implant regenerates the missing parts, whereas the limb on **D** does not regenerate and may regress severely. (After Wallace.)*

150   unirradiated Schwann sheath cells and a few fibroblasts from around the implanted axons apparently undergo mitosis and give rise to the whole blastema and to the subsequent regenerated limb. This is demonstrated by implanting nerve from a pigmented genotype into a nonpigmented host limb; the tissues of the regenerated limb are pigmented as they would be in a donor. In other experiments, it has been shown that the neurotrophic effect of nerves is not eliminated by x-irradiation so long as the axons in a limb remain connected to their nerve cell bodies. Regeneration never occurs in these experiments, however, if a source of unirradiated Schwann cells (and perhaps fibroblasts) is not present. In other words, the neurotrophic action of nerve is not sufficient to support regeneration; unirradiated Schwann cells appear to be essential instead, in this experiment. It is surprising enough that these cells can give rise to the various mesenchymal cell types—bone, muscle, and so on—but what is truly amazing is that the shape of those organs is correct and that a functional limb results! Clearly, the genetic information must be available for use in the unirradiated cells, and the equivalent of a progress zone may be at work!

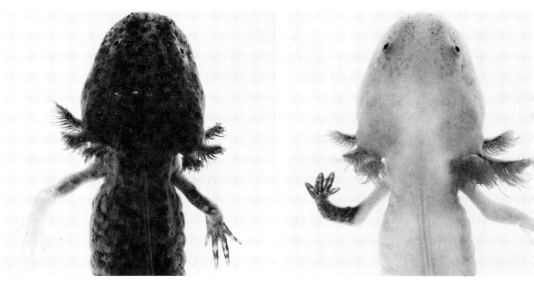

Figure 11.5

The left photograph shows dark, Dd-genotype axolotl (a kind of salamander) whose left arm was irradiated (so that regeneration could not occur) and amputated distal to an implant of a "white" dd-genotype brachial nerve. Note the white arm regenerate. The right photograph shows a white dd animal whose left arm had been irradiated and amputated just distal to an implant of a "dark" Dd-genotype nerve. The regenerate is dark. (Courtesy of H. Wallace and the Journal of Embryology and Experimental Morphology.)

In summary, both a neurotrophic action and possible cell donation may be involved in how the action of nerves leads to limb regeneration. In the normal, unirradiated condition, the major source of blastema cells is probably the mesenchymal tissues near the amputation site. Nevertheless, the Schwann cells and fibroblasts of nerve trunks may make contributions to the blastema.

How do these results relate to nerves in embryos? Only scattered observations have been made, but increased rates of RNA, DNA, and protein syntheses have been measured when nerves and other tissues are intermixed. Thick nerve trunks or ganglia (aggregations of nerve cell bodies of the peripheral nervous system) are often found at sites where morphogenesis or differentiation is about to begin. For instance, thick bundles of axons extend to a series of sites on each upper lip of mouse embryos, just at the time when development of a large vibrissa or "whisker" hair commences at each such site. We do not understand how nerves arrive at such specific sites so early in development, nor whether they play a role in stimulating mitosis, morphogenesis, or differentiation once they are there. This is a subject ripe for investigation.

Differentiation and Nerves

We will now turn to a second major role of nerves in development, one in which the cell phenotype is at stake. Two tissue types, skeletal muscle and taste buds, warrant discussion.

The basic contraction characteristics of mammalian muscle depend on the kind of nerve innervating that muscle cell. If we interchange the nerve supplies to "slow" and to "fast" muscles, we observe that the "slow" muscle speeds up its rate of contraction, whereas the "fast" muscle slows down. The consequences for the muscle cells are basic, since even the kinds of soluble proteins are altered as a result of switching the nerve inputs. For example, let's consider lactic dehydrogenase (LDH), an enzyme whose function depends on the kind of polypeptide subunits present. "Slow" muscle normally possesses mostly "heart"-type subunits that are compatible with oxidative metabolism. "Fast" muscle normally contains "muscle"-type subunits that are associated with glycolytic pathways. When a "fast" nerve is caused to innervate a "slow" muscle, there is a shift in LDH subunits so that "muscle" ones predominate. These and other measurements indicate that the types of enzymes and biochemical processes in muscle cells can be controlled by the type of nerve present.

The *distribution* of the chemicals that are essential to normal nerve-muscle interaction also depends on the presence of nerve (see Figure

152

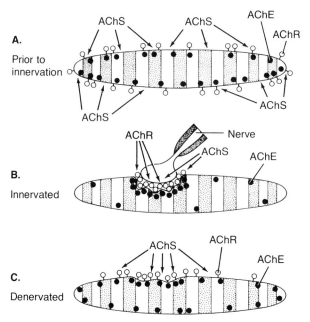

Figure 11.6

Nerve-muscle interaction and motor end plate properties.
A. In many developing muscle cells, sensitivity to acetyl choline (AChS) appears and is spread over the whole surface prior to innervation (though, a few "hot" spots of high sensitivity may be detected). Receptors for acetyl choline (AChR) are also spread on the surface. The enzyme acetyl cholinesterase (AChE) is synthesized and distributed through the cytoplasm. B. After innervation and construction of the motor end plate, AChR are essentially restricted to the motor end plate, so that sensitivity to acetyl choline is localized there too. The great bulk of AChE activity is at the muscle membrane of the end plate, though low activity can be detected elsewhere in some muscle cells.
C. Following destruction of the nerve that innervates the muscle cell, AChE activity drops and spreads, AChR appear laterally from the degenerating motor end plate, and sensitivity to the neurotransmitter ACh spreads in a similar manner. In some muscles, drugs blocking ACh release from intact nerves produce the same kinds of alterations. (See Gutmann, Harris, for literature.)

11.6). When muscle cells first develop, they are sensitive to the neurotransmitter acetylcholine over their whole surface. Once a nerve innervates a muscle cell, that sensitivity becomes restricted to the motor "end plate" region (the site in the adult where the nerve normally releases acetylcholine to cause muscle contraction). If a nerve going to a muscle is cut, sensitivity to acetylcholine spreads over the surface

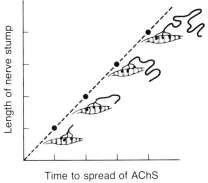

Length of nerve stump

Time to spread of AChS

Figure 11.7

An experiment showing the effect of trophic factor on motor end plate properties. Imagine that we cut the nerve axons leading to four muscle cells at progressively longer distances from the muscle cells. If we then monitor any of the major properties of the motor end plates, say, sensitivity to applied acetylcholine (AChS), then we might find the relationship described by the line in the graph. Specifically, the longer the stump, the longer it takes for the degenerative change to start (spread of AChS). This can be interpreted as meaning that trophic substances move slowly down the axon, are released at the motor end plate, and keep it and the muscle cell in its normal differentiated condition. When the hypothetical flow and release ceases, the sensitivity spreads. Hence, the longer the stump of nerve, the more time it takes for the supply of the putative trophic factor to be depleted. (Note: a precise linear relationship, as shown, has not been established; the general relationship is correct, however.) (See Gutmann, Harris, for literature.)

once more, indicating that nerve plays a continuing role in the maintenance of the differentiated condition.

Besides sensitivity to the neurotransmitter, the motor end plate is also characterized by the presence of acetyl cholinesterase, the enzyme that rapidly hydrolyzes the neurotransmitter after its release from the nerve ending. This enzyme does not accumulate at motor end plates in the absence of nerves.

It is not known how nerves exert these various effects on muscle cells. On the one hand, molecular exchange (other than neurotransmitters) is possible between the nerve ending and the responding cell. Lentz has shown, for instance, that an extract from nervous tissue has an important effect on muscle cells in culture. If the nerve leading to a cultured amphibian muscle is cut, then acetyl cholinesterase activity falls rapidly. If a nerve extract is included in the fluid nutrient medium of the culture, then the enzyme activity decreases much less rapidly. Equivalent observations on cultured chick muscle cells have shown that some of the very large molecules (greater than 300,000 daltons) derived from nervous tissue support protein synthesis, acetyl cholinesterase activity, and the state of differentiation of skeletal muscle cells. Perhaps these experiments, and those of Singer on the trophic molecules

154   that stimulate protein synthesis in denervated limb regenerates, will finally lead us to the elusive trophic factors by which nerves may support the differentiated state in muscle cells.

On the other hand, some neurobiologists are convinced that it is the frequency of muscle contraction elicited by nerve (the work done, etc.) that controls muscle biochemistry. Thus, "tonic" nerves, which fire repeatedly at low frequencies, evoke the characteristic prolonged contractions of "slow" muscle; "phasic" nerves fire short, high-frequency discharges that are responsible for "fast" muscle contractions. Obviously, the "work" patterns of the two muscles differ because of the way they are stimulated; the biochemical machinery of the muscle cells is adapted to the differing contraction patterns, and so indirectly, to the kind of nerve present. If this alternative proves to be the correct explanation of the phenomenon, then either of two possibilities will apply to the nerve-muscle interaction: first, the interaction might best be regarded as a case of adult-type functional interaction between cells; as such, it would not be of general interest for embryonic tissue interactions. The second possibility, as pointed out by W.J. Dickinson, is that other embryonic tissue interactions are side-effects of normal cell function: the muscle cell has certain properties because it "works" in response to nerve cell function; so too could some embryonic cells "work" in response to changing properties of neighboring cell populations and thereby alter their own development course. If this second possibility proves likely, then the nerve-muscle model will have general interest. Furthermore, if future experiments prove that molecular exchange of trophic factors takes place between nerve and muscle (as outlined above), then we will have the opportunity to investigate a true cell-to-cell interaction, not the usual tissue-to-tissue interaction we have been dealing with in this book.

Muscle Effects on Nerves

Motor nerves of both the central and the peripheral nervous systems depend on proper connections with end organs for survival. In embryos, a surplus of motor neurons develops in the motor horn portions of the spinal cord (the ventro-lateral quadrants) and in parasympathetic ganglia. All of these motor cells apparently send axons toward their end organs (limb muscles, eye muscles, etc.), but only a fraction succeed in establishing motor end plates. Once neuromuscular connections are set up by some cells, axonal tips arriving later seem to be excluded from forming synapses with the same muscle cells. The

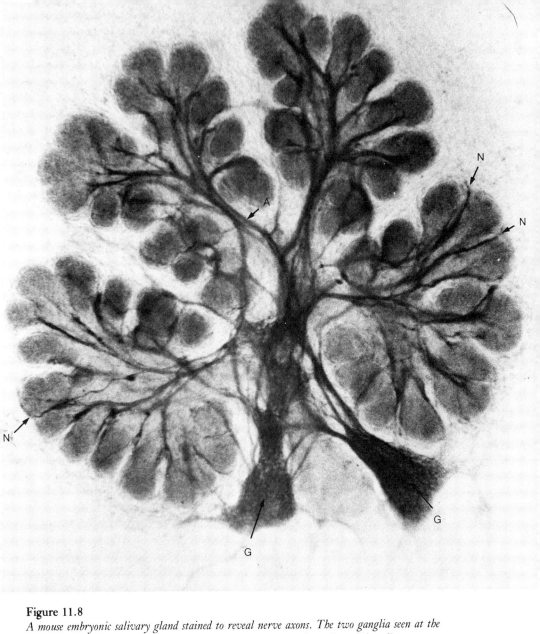

Figure 11.8

A mouse embryonic salivary gland stained to reveal nerve axons. The two ganglia seen at the bottom are the sites of the nerve cell bodies. Axons grow over the surface of the salivary epithelium as it undergoes branching morphogenesis (described in earlier chapters). Note the remarkable conformity of the axons to the epithelial shape; often bundles of axons are associated with each of the clefts that appear on the epithelium. The axons do not grow outward into the surrounding mesenchyme or cover the epithelium in random fashion, and then sort out secondarily after epithelial morphology is complete. Instead, the morphogenesis of the nerves and that of the epithelium that will ultimately be served by those nerves go on concurrently. This is one of a number of cases that suggest that the solid substratum upon which nerve axons grow serves as a guide to the growth cones. Thus the shape and distribution of nerve axons in space—a crucial element of nerve networks—is governed by the environment of the nerve cell, not by its own intrinsic information. (Courtesy of M.D. Coughlin, Develop. Biol., 43 (1975), 123.)

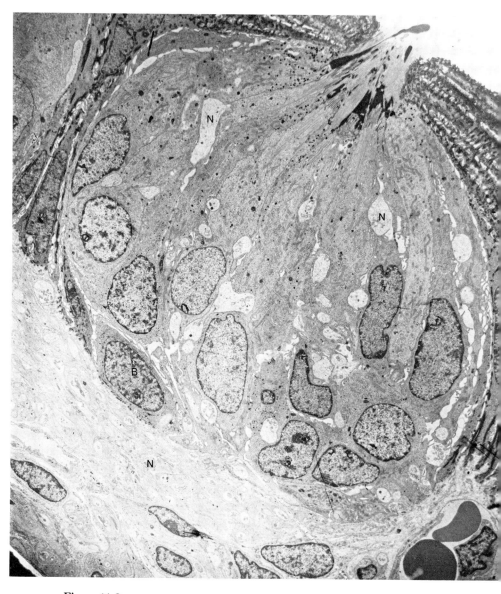

Figure 11.9

A section through a taste bud in the tongue of a rabbit (this one is located on a folliate papilla; see Figure 11.14). About 30 to 80 cells make up a bud of this type. The opening at the top of the bud is the site where substances in the oral cavity reach the taste receptor cells. The regions that here appear light gray beneath the bud and between some of the taste-receptor cells are sections of nerve dendrites that will carry impulses toward the ganglion and brain (via cranial nerve IX; see text). The cell marked B is probably a basal cell, one that divides and gives rise to new taste receptor cells in the bud. (Courtesy of R.G. Murray, from I. Friedman, ed., The Ultrastructure of Sensory Organs, *American Elsevier, 1973.)*

unsuccessful neurons become necrotic and are destroyed. This "normal" cell death in embryos occurs at about the same time that the surviving cells are acquiring their functional connections with end organs.

A specific example makes these relationships clear: about 6,900 motor neurons differentiate structurally and functionally in a ciliary ganglion of a chick embryo (these ganglia are located near the site where the optic nerve exits from the eyeball). Then, between the ninth and thirteenth days of incubation, about half these cells become connected to the intraocular muscles (of the iris and the ocular blood vessels). The other half die, so that only about 3,400 neurons are present in a ciliary ganglion of an adult bird. If an eye is removed from a young embryo so that the normal connections cannot be made, then over 90 percent of the original 6,900 cells die. An equivalent massive death occurs in a motor horn if a limb is missing.

Now let us consider the opposite situation, where extra end organs are present. Polydactyly (Chapter 6) involves extra digital bones and muscles being present; there is a corresponding increase in the number of nerve cells at that level of the spinal cord. Similarly, following transplantation of an extra limb onto the flank of a host embryo, muscles develop and become innervated. "Excess" neurons survive and differentiate.

These cases lead to the hypothesis that end organs provide molecules to the nerve ending; these substances may be transported back through the axon to the nerve cell body, where they help to maintain viability. Though experiments using radioactive precursors of protein have shown such transfer and transit, the identity of the molecules has not been established, nor is it known whether the labeled molecules are responsible for neuronal survival.

Interaction between Sensory Nerves and End Organs

Next, let's turn to interactions between a sensory organ and nerves. Taste buds on the tongue of most vertebrates consist of local collections of epithelial cells that are specialized on their outer surfaces to respond to foreign molecules and initiate the sensation of "taste." Nerve processes abut on the inner surfaces of the taste cells and carry impulses away from the tongue toward the brain (Figure 11.10).

Taste cells, like others associated with epithelia on the body, have a finite life; after a certain period (about 10 days), a given cell will die and be replaced by a new cell that arises by mitosis from an adjacent "germinal" cell in the tongue epithelium. Thus differentiation of new taste cells is a normal part of adult life.

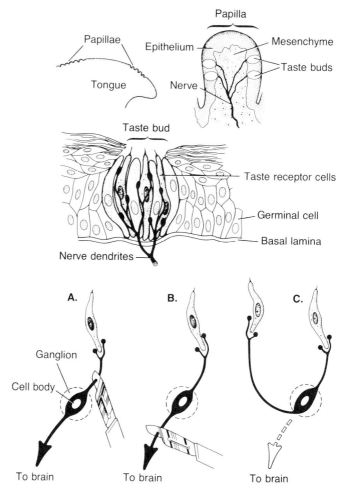

Figure 11.10

*Taste buds are located on various types of papillae found on the tongue. Nerve processes penetrate the basal lamina, so that when taste receptor cells are activated by salt, sweet, etc., impulses travel over the processes toward the brain. The diagrammatic representations show that the sensory nerve cell bodies are located in a ganglion. Severing the nerve processes between taste bud and ganglion (**A**) results in a loss of taste receptor cells. The same operation between ganglion and brain (**B**) does not affect the taste cells. In a different type of experiment (**C**), the branch of a taste receptor nerve that normally carries impulses from the ganglion to the brain is diverted to the tongue. Taste buds develop, showing that the effect of nerve in causing taste bud formation can originate from either "end" of a taste nerve cell. Unidirectional transport from the cell body of the normal nerve toward the tongue seems unlikely. (See A.A. Zalewski, Exptl. Neurol., 25 (1969), 429.)*

The integrity of taste buds depends on nerves. If the gustatory nerves are cut between the sensory cell bodies and the buds, then the bud cells that are already present degenerate and are lost and new ones do not reappear (interestingly, the tongue epithelium itself also becomes unusually thin in the absence of the nerves). When the gustatory nerves are cut between the sensory-cell bodies and the brain (Figure 11.10), there is no effect on taste buds. Thus the nerve cells do not have to remain connected to the central nervous system in order to keep taste buds intact.

The interaction between nerve and taste bud seems to be specific. Suppose we remove the sensory nerves that lead to the tongue epithelium and permit all the taste buds to degenerate and be sloughed away. If sensory nerve bundles that would normally connect to chemoreceptors are brought back into the vicinity of the tongue, reinnervation occurs, and new taste bud cells appear. If a pure "motor" nerve, the hypoglossal, is substituted instead, no taste buds form. Similarly, if nontaste sensory nerves are substituted, there is no taste-bud development, even though such nerves grow endings into the tongue that become functionally active (they sense chilling or touch, for example). Thus the interaction is quite specific: only chemoreceptor nerves elicit taste bud formation.

The relationship between sensory nerves and responding system is thus considerably more complex than the similar relationship for limb regeneration, where mass (or cross-sectional area) appeared to be a crucial factor. This difference is emphasized further by cross-innervating tongue epithelium, so that the tiny chorda tympani taste nerve must serve the many taste buds originally innervated by cranial nerve IX. Since the much smaller nerve can support those buds, it seems likely that quality, not quantity, is the key to the interaction.

The mechanism by which certain sensory nerves act on the tongue epithelium is not known. Appearance of taste cells begins only after nerve axons have penetrated the basal lamina beneath the epithelium and have established intimate contact with the epithelial cell surfaces. We do not know, however, what mechanism is involved in the ensuing interaction. Among the possibilities are: (1) exchange of trophic molecules; (2) electrophysiological effects arising from "antidromic" activity in the sensory nerve neurites (i.e., impulse traffic moving toward the taste cells); or (3) cell surface interactions that depend on GAGs, glycoproteins (a class of protein-sugar complexes; see Chapter 15), or molecules that are components of the plasma membrane itself (intramembranous particles; see Chapters 14 and 16).

160 Sensory End Organ Effects on Nerves

The kind of dependence of nerve on muscle that we discussed earlier in this chapter also operates in the sensory system. Removal of an embryonic limb bud results in reduced size of the sensory ganglia that would normally have served the missing limb. Similarly, removal of the developing eye in a chick or in other vertebrate embryos leads to excessive cell death in the optic tectum portion of the brain, the site where axons from the neural retina would normally have formed synapses. Just as with the motor nerve and muscle system, the actual proliferation of neuroblasts and initial phases of differentiation do not seem to depend on the sensory end organs. Maintenance or survival is the sensitive process.

A protein called Nerve Growth Factor (NGF) provides a model for this maintenance function. NGF is essential for development of spinal ganglion sensory neurons and for sympathetic neurons. In an experimental situation, such cells extend long neurites and remain in a healthy condition. Sensory cells are dependent on the protein during only a limited period of embryonic development, and, interestingly, it is only during this phase that receptor molecules for NGF can be detected on the surface of cells. Sympathetic cells are dependent on NGF for a longer time, but it is not known whether the protein must be present for adult cells to survive. If antibodies to NGF are injected into developing embryonic or newborn mammals, sympathetic cells die, presumably because the antibody-NGF complex prevents the normal action of NGF.

The site of manufacture of NGF is not proved, though much is stored in rodent salivary glands (in males or in females that have been

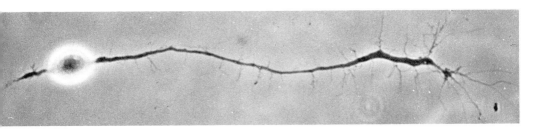

Figure 11.11
A neuron from a chick embryo's dorsal root ganglion growing in culture in response to nerve growth factor. Axons like the one seen here may branch many times in culture; the number of branches depends on properties of the substratum on which the growth cone moves. Similar cells from younger embryos do not require nerve growth factor in the medium to survive and to construct axons.

injected with testosterone). Small amounts of circulating NGF have been measured in some organisms; so there may be some systemic dissemination. More interesting, perhaps, is the suggestion that NGF may be transferred to sympathetic cell nerve terminals from end organs. If radioactively labeled NGF is injected into the cavity of an eye, some NGF appears to be picked up by nerve cells and transmitted back through axons to sympathetic nerve cell bodies. Perhaps the various tissues in the body innervated by sympathetic nerves produce NGF, and transfer it to nerve endings to permit a similar transport process to go on. Many careful experiments must be done before these relationships are proved; nevertheless, they provide a model for the way that sensory or motor end organs could act back on nerves to promote survival.

General Features of Nerve-Based and Hormone-Based Interactions

Several important lessons will become apparent if we compare nerve with hormone actions on responding organs. Let us consider the three major expressive processes: (1) mitosis; (2) differentiation and morphogenesis; and (3) maintenance of functioning phenotype.

These steps are seen in Figure 11.12, and generalized conclusions about them, for each of the interactions, are listed in Table 11.1. First, note that estrogen appears to act on all three processes in the oviduct; it is both a trigger of development and a regulator of ovalbumin synthesis in the differentiated cells.

The nerve-muscle interaction is in striking contrast to this broad action. Both prospective muscle cells (myoblasts) and nerve cells (neuroblasts) proliferate initially, and undergo differentiation and morphogenesis, independently of what the other cell type is doing. Then, after innervation, groups of muscle cells become segregated into individual muscle organs (gastrocnemius, biceps, etc.); sensitivity to neurotransmitter becomes restricted; motor end plate biochemistry matures; and so forth. As we saw, these final aspects of phenotype depend on continuing innervation, just as ovalbumin synthesis is a function of estrogen concentration.

What about the reciprocal effect of muscle on nerve? The experiments certainly emphasize that there is a discrete period when nerve and muscle must be connected or neurons will die. But it is not correct to extrapolate from that to an adult and to assume a maintenance function of muscle on nerve. Recall, for instance, stories of "phantom pain"

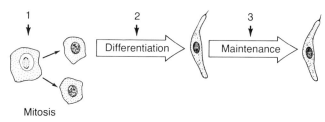

Figure 11.12
Possible times when nerves may act on taste buds: (1) during generation of the cells; (2) during differentiation; or (3) during the mature, functional state.

TABLE 11.1
Developmental responses to interactions between end organs, nerves, and hormone

Process	Interaction	Conclusion
1. Mitosis	nerve on taste bud	dependence proved
	taste bud on nerve	no evidence favors
	nerve on muscle	no evidence favors
	muscle on nerve	no evidence favors
	estrogen on oviduct	dependence very likely
2. Differentiation and morphogenesis	nerve on taste bud	dependence possible
	taste bud on nerve	no evidence favors[a]
	nerve on muscle	some aspects are dependent
	muscle on nerve	some aspects may be dependent[a]
	estrogen on oviduct	dependence very likely
3. Maintenance	nerve on taste bud	not clear; for other sense organs, some dependence is possible
	taste bud on nerve	not for survival; minor aspects of phenotype possible[a]
	nerve on muscle	atrophy without nerve
	muscle on nerve	not for survival; minor aspects of phenotype possible[a]
	estrogen on oviduct	dependence very likely

[a]*Some experiments indicate that the "wiring" pattern of the nervous system may be dependent upon peripheral connections.*

in human limb amputees—pains, itching, and other sensations "in" the missing limb. That kind of phenomenon, the ability of severed motor or sensory nerves to regenerate neurites, and other observations, all suggest that nerve survival after the critical embryonic period is not dependent on muscle.

What of the seemingly clearcut case of nerve and taste bud? Careful examination of the experiments shows that it has not been proved that nerve acts *both* on mitosis and on differentiation. With nerve present, both occur. Perhaps nerve is only a mitotic trigger for preprogrammed cells, ones which carry out their own developmental timetable, but which because of our experimental design, appear to be dependent on nerve for differentiation. To carry the argument further, we must consider the fact that the differentiated cells in a taste bud survive after the nerve is cut for about the same time that they would have as in-nervated cells. Absence of nerve does not shorten their lifespan; it leads to an absence of new taste receptor cells to replace the older, dying ones. Thus, there is little hint of a maintenance effect of nerve when we speak of cell-level events. On the other hand, past investigators have correctly pointed out that taste buds, as organs, are not main-tained in a denervated tongue.

We are not merely sharpening Occam's razor on the nervous sys-tem or retreating into semantics. The student will recognize the same difficulties in logic that we encountered previously in analyzing more complex tissue interactions. Perhaps the strongest message is that future workers should be careful to distinguish between the multiple responses of cells or tissues in interactions, so that primary causal events can be identified.

Stem Cells, Taste Buds, and Nerves

Figure 11.13 poses another significant question about the source of taste bud cells and the concept of permissive versus instructive inter-actions. Is there a restricted, germinal, "stem" cell line in tongue epi-thelium that awaits the stimulus of nerve in order to produce taste bud cells? If so, a separate stem group of cells must give rise to the sur-rounding keratinizing populations. Or do the germinal cells of tongue epithelium have the capacity to produce daughter cells of either the keratinizing or the taste bud type? If so, then the nerve must say "di-vide and form a taste receptor cell"—this would be an instructive interaction. If the first possibility is correct, then the nerve merely says "divide" to cells that are already restricted to forming taste receptors.

Experiments by Oakley show that the local tongue environment governs the type and distribution of taste buds (see Figure 11.14). Thus, in various cross-innervation situations, the taste buds that form are appropriate to the region of the tongue, not to the type of taste

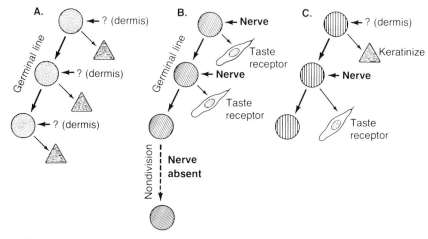

Figure 11.13

*Stem cell lines in the tongue. One alternative is that two germinal stem lines are present: **A** responds to the dermis-like mesenchyme of the tongue and forms keratin; **B** responds to nerves by dividing and producing taste receptor cells. The second major possibility is that a single stem line **(C)** is present: if mesenchyme acts, daughter cells keratinize; if nerve acts, daughter cells form taste receptors.*

Figure 11.14

*Tissue environment controls the type of taste bud. In the normal rat tongue **(A)**, cranial nerve IX innervates the posterior folliate papillae, each of which bears many taste buds, each with a characteristic sensitivity. The* chorda tympani *innervates the anteriorly placed fungiform papillae, each of which has a single taste bud that is different in character from folliate taste buds. If the nerves are cut, the original taste buds disappear. Then, if the nerves are crossed and allowed to reinnervate the wrong portion of the tongue **(B)**, taste buds develop again. Multiple buds form in the rear, singles in the front, just as they did with the original nerve supply. It seems likely, therefore, that the type of bud is governed by the local environment of each type of papilla, not by the specific nerve present. (See Oakley for references.)*

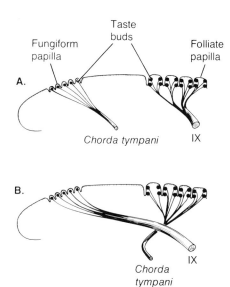

nerve that is present. The local epithelial and mesenchymal populations seem limited in the way they can respond to what appears to be
a nonspecific stimulus by taste nerves. Unfortunately, the experiments
do not address the stem cell question outlined above, since the local
epitheliomesenchymal populations might be restricted to forming only
certain types of taste buds, but the germinal cells still might require
a specific instruction about whether to form taste buds or keratin.

Despite these lingering uncertainties, little evidence favors the
concept that nerves act on restriction and determination. If anything,
the rule is that the nervous system develops by itself, end organs, such
as muscle, do too, and interdependence only appears secondarily.
The exceptions, of course, are mainly in the sensory system, where
only further work will tell us whether nerves act instructively. Whatever the outcome of such studies, there can be no doubt that the nervous system must be included with the endocrine system as a source
of important information for the expressive phase of development.

CONCEPTS

A minimal quantity of nerve tissue is required for the initial stages
of amphibian limb regeneration.

Many responses occur if there is enough nerve tissue in an amputated limb, including mitosis, GAG production and degradation,
and formation of the regeneration blastema.

Some important characteristics of differentiated skeletal muscle
cells are controlled by the type of nerve present.

During embryonic development, motor nerves pass through a
period in which they must be in contact with muscle in order to survive.

Some sensory organs depend on innervation for integrity.

The specificity of interaction between nerve and sensory organ
is strict: only taste nerves can support development of taste buds; other
sensory nerves cannot.

During embryonic (or even postnatal) development, sensory
nerves pass through a period in which they must be in contact with
end organs in order to develop or to survive.

166   It is not clear that either nerves or hormones act instructively on restriction or determination.

The degree of restriction of stem cells in epithelia is still undefined.

REFERENCES

General:

M. Jacobson. 1970. *Developmental Neurobiology.* Holt, Rinehart, Winston. The best source of literature in this field up to 1969. The text is spotty in treatment of some subjects, and takes strong positions on some, but is useful, particularly on nerve-end organ interactions.

A.J. Harris. 1974. *Ann. Rev. Physiol.*, *36*, 251. A more recent discussion of muscle and sense organ interactions with nerves.

E. Gutmann. 1976. *Ann. Rev. Physiol.*, *38*, 177. The latest literature on nerve-muscle relationships in development.

GAG, hyaluronate, hyaluronidase, nerves, and limb regeneration:

G.N. Smith *et al.* 1975. *Develop. Biol.*, *43*, 221. First-class science, critical interpretation, and good literature on nerves in relation to the GAG role in development.

Muscle activity and development:

R.J. Tomanek. 1975. *Develop. Biol.*, *42*,, 305. Discussion of nerve-muscle interaction in terms of work, functional demands, and hyperplasia, not just trophic factors. See also: A. Pestronk *et al.*, 1976. *Nature, 260*, 352; and, P. A. Lavoie *et al.*, 1976. *Nature, 260*, 350.

Cell death in ciliary ganglia:

L. Landmesser and G. Pilar. 1974. *J. Physiol.*, *241*, 715. See also *J. Cell Biol.*, *68*, 339 and 357. Papers describing cell death in the presence and absence of end organs.

Sheath cells and limb regeneration:

H. Wallace. 1972. *J. Embryol. Exptl. Morphol.*, *28*, 419. Irradiation and nerve-implant experiments that imply Schwann cell contribution to the blastema.

Taste buds:

B. Oakley. 1974. *Brain Res.*, *75*, 85. This paper shows how mesoderm controls the types of taste buds formed. It is an excellent source of references and a good summation of the major problems in the field.

Nerves and blood vessels:

A.R. Smith and L. Wolpert. 1975. *Nature 257*, 224. This brief, but provocative, paper is an entry to a small but important field.

Chapter Twelve:

Prometheus chained for eternity to Mount Caucasus suffers for his treachery. The vulture of the Immortals feeds on his liver each day, but whatever is consumed regenerates during the night. Kiortsis and Trampusch consider this myth to be the oldest report on regeneration. Atlas staggers near Prometheus, bearing the world on his shoulders. This photograph is of a kylix, or drinking cup, from Caere, Sparta, and consists of black figures on a red background. The kylix dates from about 550 B.C. and is about 20 cm in diameter. (Vatican Museum, 1298; Alinari, 35838. See A. Lane, Greek Pottery (London: Faber and Faber, 1948).

Inhibitory Interactions

"Pluck out my lens again and again, but I will see you yet, my friend," said the salamander.

A peculiar feature of certain adult tissues is that inhibitory factors may influence stability of cell phenotype or rates of mitosis. The eye of a salamander provides a striking example of how tissue stability depends on an inhibitory stimulus. If the lens is removed from an adult salamander's eye, the nearby iris produces a new lens. Iris tissue normally consists predominantly of pigment cells and of myoepithelial cells (epithelial cells that possess muscle-like contractile machinery which functions to close the iris opening). In the days following lens removal, pigment cells located on the posterior surface of the dorsal iris begin to lose pigment granules (Figure 12.1). This loss of phenotype is accompanied by an appearance of mitotic activity. Gradually an increasingly large mass of unpigmented cells builds up. The cells undergo morphogenesis, forming a typical lens epithelium and set of lens fibers. Ultimately, a normal functional lens is produced. Then the dorsal iris population ceases mitosis and develops the pigment phenotype once more.

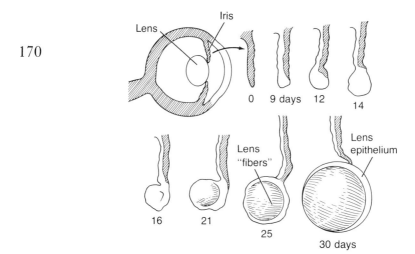

Figure 12.1
*Stages in lens "regeneration" from the dorsal iris in a
salamander's eye. Depigmentation, mitosis, and differentiation
into lens fiber cells characterize this remarkable transition
in cell type. (See R.W. Reyer, Quart. Rev. Biol. 29
(1954), 1.)*

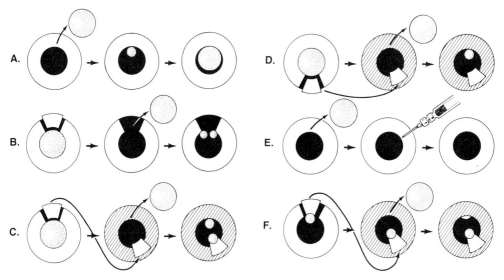

Figure 12.2
*The fascinating world of lens regeneration. **A**. Control: lens out; dorsal iris regenerates a lens. **B**. Dorsal iris out;
lens out; two lenses regenerate from edges of iris nearest the dorsal sector. **C**. Dorsal iris transplanted to lower
quadrant host eye; lens out of host eye; two lenses form in synchrony. **D**. Ventral iris transplanted to lower quadrant host eye; lens out of host eye; one lens forms from host dorsal iris.
The ventral iris lacks ability to form lens. **E**. Lens out; inject daily with fluid (aqueous
humor) from a normal eye containing a lens; no lens regeneration until injections are halted.
F. Dorsal iris with attached regenerating lens transplanted to lower quadrant host eye; lens out of host eye;
implanted regenerate continues to develop; no lens forms on dorsal iris of host eye (or a small one starts to form and
regresses). This inhibitory effect is only seen when the implanted regenerate has reached a specific stage of development. (Experiments by L.S. Stone. For references, see R.W. Reyer, Quart. Rev. Biol., 29 (1954), 1.)*

A number of the "rules" of cell development we have discussed are obeyed in this example. Differentiative phenotype and rapid mitosis are inversely correlated. Mitosis (DNA synthesis) intervenes between the two discrete cell states (pigment, lens phenotypes). However, is the iris-to-lens transition a contradiction to the concept of *determination?* Strictly speaking, it seems so. But is it possible that, for unknown reasons, cells of certain types of tissues in the eye retain the capacity for a limited number of alternative forms of development? (Recall the possibility that germinal epithelial cells of the tongue may produce either keratinizing or taste receptor daughter cells.) It is known, for instance, that both in salamanders and in chick embryos, pigmented retinal cells can give rise to a new neural retina (the cell layers where light is absorbed, nerve impulses are generated, and initial routing of impulses occurs). Furthermore, in salamanders these "regenerated" retinas can hook up correctly with the brain so that normal vision is restored. In these cases of "alternative gene usage," no evidence suggests that cells of the iris, retina, oral epithelium, and so forth, can undergo such a radical transformation that they can synthesize hemoglobin, digestive enzymes, or other proteins characteristic of widely divergent cell types. Therefore we must conclude that, at least in the eye system and in certain epithelial systems of some higher animals, there is a limited repertoire of alternative cell types that can be formed. If so, what ensures phenotypic stability under normal circumstances?

That is where an "inhibitor" may play a role. First, let's remove the lens from an eye. If we then insert a lens back into the same eye, the dorsal iris remains pigmented and no new lens forms. The inhibitory activity shown by the inserted lens probably originates from its differentiated cells. Thus, primitive undifferentiated lenses cannot inhibit the iris-to-lens switch. Only mature lenses act that way. The identity of the inhibiting agents is unknown. Separation of lens proteins by starch gel electrophoresis has been attempted. If pieces of gel containing two of the lens proteins are placed in eyes from which the lenses have been removed, no regeneration occurs; the iris remains inhibited. These experiments imply that in a normal eye an inhibitory substance (or substances) emanates from the lens and resides in the fluid of the anterior eye chambers. Those fluids bathe the iris tissue and presumably hold its cells in check, maintaining them as iris, and preventing them from forming lens.

What is so intriguing and perplexing about this case is the difficulty in imagining any specific forces of natural selection that would lead to the evolution of a special system of inhibitory interactions just

172   for eye tissues. One doubts, for instance, that many lenses have been lost in nature without the loss of the head as well! Consequently, we must keep the eye system in mind as we think about the sources of phenotype stability for other adult cell types, and as we construct models for the ways in which tissues can interact in embryos.

Inhibitors in Adult Tissues

Inhibitory factors may also operate in controlling tissue or organ size in adult organisms. Here we are not dealing with changes in cell type, but with control of cell population size. A simple case occurs in adult surface epidermis, a tissue in which continual mitotic activity in the basal, "germinal" layer is perfectly coupled to the rate at which dead, keratinized cells are shed from the surface of the epidermis. If you repeatedly touch a piece of adhesive tape to the same region of your skin, the tape will remove a significant number of those outer, differentiated cell layers (see Figure 12.3). A few hours later, in the basal layer beneath the thin region, increased mitotic activity will start and daughter cells will be produced at a greater than normal rate until the gap is filled. Equivalent responses are seen in other kinds of wounds. The system behaves as if a negative feedback control loop were operating, with the differentiated keratinized population being the source of a signal that inhibits or limits mitosis. A decrease in the number of differentiated cells is translated into a decreased inhibitory signal, and the system corrects itself by more rapid mitosis.

A glycoprotein (sugar-protein complex; Chapter 15) that has been isolated from extracts of epidermis seems to act as a specific inhibitor of mitosis in epidermal cells. The substance, called a "chalone" (meaning "to make slack," i.e., to slow down), is not species-specific, since mouse factor will inhibit human epidermal mitosis, but is tissue-specific, since the epidermal chalone has no effect on other epithelia and their chalones have no effect on epidermal mitosis. These and other observations lead Houck and Attallah to list the properties of any putative chalone as:

1. Cell specificity (i.e., acts only on the same type of cell that it originates from).

2. Endogenous origin (i.e., is produced by the type of cell that it affects).

3. Lack of species specificity.

4. Reversibility or lack of cytotoxicity (the putative chalone should not function merely as a toxic agent, rendering cells unhealthy so that they do not divide).

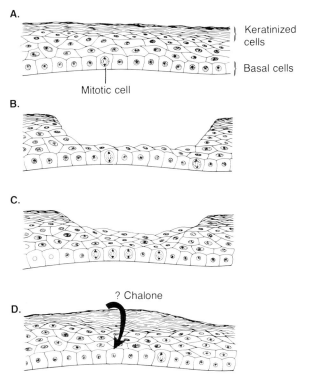

Figure 12.3

Compensatory growth in adult epidermis. In normal epidermis, a low frequency of mitosis is coupled to a low rate of loss of keratinized cells from the surface of the skin. If a patch of the differentiated keratinized cells is removed (B), mitosis speeds in the underlying basal region (C), and new daughter cells are generated to fill the gap. When the wound is repaired (D), mitosis returns to normal in the underlying basal population. It is hypothesized that a substance that originates in the differentiated cells acts back on the germinal basal cells to inhibit mitosis there. That regulator molecule is called a "chalone."

With these criteria in mind, let us turn from the epidermis, where mitosis is common even without an injury, to certain internal organs with little mitosis, where the possibility of negative feedback is again encountered. If a kidney is removed from an adult mammal, the remaining kidney increases substantially in size. In part, this increase results from a burst of cell division activity in the remaining kidney, which begins about 48 hours after the operation. Even earlier a substantial increase in RNA content occurs, largely because the normal degradation (turnover) of ribosomes and ribosomal RNA ceases. The

174   final enlarged kidney carries out substantially more filtration of blood
in its glomeruli, and so helps to compensate for its missing partner.

A chalone may be involved in this phenomenon. A typical assay
system consists of cultures of the dorsal body of a *Xenopus* tadpole,
including kidney and skin tissues. Mitosis and DNA synthesis slow
down in the developing kidney tissue if an extract of adult kidney is
added (prepared like the epidermal chalone) to the medium. Epidermis
on the same explant shows no inhibition. Other experiments imply
that the kidney chalone is present in serum, and so circulates everywhere in the organism. Studies on this and other chalones show little
evidence of toxicity when they are tested by injections into whole
organisms or applied directly to cells in culture.

By far the largest volume of work in this area is on the mammalian
liver. If about 75 percent of the liver is removed, the remaining liver
lobes increase spectacularly in size, to approximate the original mass
of the liver. In so doing, the remaining liver cells lose their functional
phenotype, undergo mitosis, and later redifferentiate—the same sequence we saw in lens, but without the change in cell type.

These phenomena in liver and kidney have been termed "compensatory hypertrophy." (The student may encounter the term liver
"regeneration" but that is a misnomer, since the *remaining* lobes enlarge; the stumps of the removed lobes do not regenerate.) A systemic
factor is involved in the liver's compensatory hypertrophy. First, let
us graft a piece of liver beneath the skin of a host rat (genetically close
individuals are used so that immune mechanisms do not interfere with
the experiment). Several weeks later, we remove 70 percent of the host
liver and find that mitosis speeds up in the remaining host lobes and
in the implanted liver (see Figure 12.4). Only the blood vascular system could deliver a signal to initiate the latter response.

Figure 12.4
*Systemic effects on liver mitosis. In **A**, a piece of
liver is removed and implanted in a host rat
(**B**). Then a major portion of the host's liver is
removed. Mitosis and other aspects of the compensatory hypertrophy phenomenon are seen in
both the implant and the remainder of the host
liver. Since the effect is seen throughout the
host body, it is concluded that information
signaling compensatory hypertrophy is delivered
systemically.*

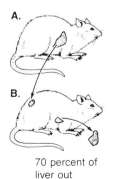

A.

B.

70 percent of
liver out

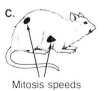

C.

Mitosis speeds

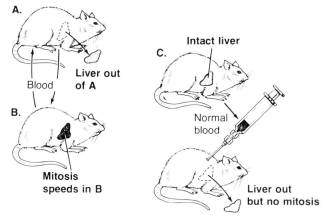

Figure 12.5

*A "parabiotic" pair of rats demonstrate that the blood vascular system is involved in liver compensatory hypertrophy. Removal of liver from rat A is followed by increased mitosis in the livers of both A and B. If blood is removed from a normal rat (**C**) and injected into a rat lacking its liver (**D**), then the normal elevation of mitosis in its remaining liver tissue does not occur. This could be explained if an inhibitor is present in normal blood (but not in blood from a rat that lacks liver tissue).*

Another line of experimentation has employed "parabiosis," a technique in which the circulatory systems of two rats are connected so that blood flows freely between the animals (Figure 12.5). If, after the circulatory systems are connected, most of the liver is removed from one of the animals, it is found that mitosis speeds not only in the operated animal, but also in its *unoperated* parabiotic mate (in the unoperated rat, there is about a sixfold increase above the level of mitosis in a control liver). Clearly, a factor in the circulatory system can influence mitosis in liver.

The parabiosis experiment does not tell us whether the removal of liver (which is a grave wounding procedure) liberates a *stimulator* of mitosis or decreases the concentration of an *inhibitor* that was present before the operation. A choice between the two possibilities has been sought in experiments in which blood from an *un*operated animal is transfused into a rat from which most of the liver has been removed. Mitosis and compensatory hypertrophy are not triggered in the animal. Similarly, if slices of liver tissue are placed in culture, DNA synthesis and mitosis are inhibited if serum from a normal rat is added to the nutrient medium, whereas serum from an hepatectomized rat stimulates DNA synthesis. The transfusion results would not be expected if a mitotic stimulant is released when liver is removed; both sets of results can be interpreted to mean that there is a mitotic inhibitor in

176 normal blood. Other researchers have isolated a small protein from serum that behaves like a liver chalone in these sorts of tests. Despite an immense investment of effort and money, the conflicting and unrepeatable results from these experiments lead one to be cautious about the liver chalone. In operational terms, however, there is little argument that the system behaves as if a negative feedback circuit were operating; that is, reduction in the mass of the liver is followed by sufficient compensatory growth to return to the original ratio of liver mass to control factor.

Though we cannot yet draw firm conclusions about the validity of the chalone hypothesis, the concept of inhibitory control of mitotic activity in a tissue is attractive and potentially important. We know nearly nothing about the control of organ and tissue size. Everyone recognizes that birds and mammals have genetically determined maximal body size. What is most puzzling is that the various body organs and tissues are found in "correct" proportion to the whole; they stop growing and retain the functionally appropriate proportion during the many years of life unless some extraordinary disease or malfunction leads to a condition of abnormality. The "correct" size of liver, kidney, or other organs is maintained despite turnover (death and replacement) of individual cells, usually at a low rate. What holds this intrinsic capacity for mitosis and growth in check?

That is where the concept of tissue-specific chalones is attractive. Is it possible that, as cells differentiate, they produce and become sensitive to agents that control their own mitosis? If so, and if the threshold for response to those agents is set at an appropriate level, a means for monitoring tissue mass would result.

In summary, we have examined cases in which the stability of cell type and of population size may be under inhibitory control. When do growth controls begin to operate in embryos? Victor Twitty exchanged developing eyes between embryos of two species of salamander that are characterized by large and small eyes. Soon the small eye increased in size more than it would have if left on the original donor embryo, and the large eye on the small-eyed host grew abnormally slowly! In both cases, the organ appears to be conforming to what is appropriate for the host environments (see Figure 12.6). Whether mechanical relations with surrounding head tissues, systemic growth regulators, or other unidentified factors account for such coordination is unknown. Nevertheless, this and many other experiments imply that controls on organ and tissue growth do operate in embryos. Only more work will identify their nature.

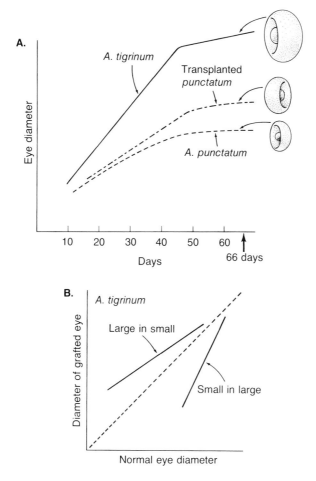

Figure 12.6
*Relative growth of transplanted eyes. In **A**, a small eye from* Ambystoma
punctatum *is implanted in the large eyed* A. tigrinum. *The implant enlarges
at its own rate at first, but on about the fortieth day, it begins to increase in
size more than it normally would; it never reaches the size of a* tigrinum *eye,
but is significantly larger than normal for its species of origin. The sketches
are of normal* tigrinum *and* punctatum *eyes and a* punctatum *eye in
a* tigrinum *host. All are drawn to the same scale as they appeared 66 days
after the transplantation in **A** was carried out. In **B**, grafts are exchanged
among tadpoles of* A. tigrinum *that vary greatly in size (because of stage
of development). "Large" eyes in small hosts grow slowly; "small" eyes in
large hosts grow more rapidly than normal. Both curves tend to approach
"unity" (dotted line) which represents the normal size through development.
(After V.C. Twitty and J.L. Schwind,* J. Exptl. Zool. 59 *(1931),* 61
and 68 *(1934),* 247.)*

178   The final, potential importance of the putative chalone-type negative feedback inhibitor is its relationship to cancerous conditions. In both solid tumors and cancers of circulating cells (such as leukemias), abnormal numbers of the cancerous cell type accumulate. In part, this results from the abnormally long lives of the cancerous cells. In addition, however, normal growth regulation may be deficient. Experiments to test whether chalones are missing or whether cancer cells are insensitive to them have yielded conflicting results. Nevertheless, the potential benefits of identifying and isolating natural growth regulator molecules justifies intensive work in this area of biology.

CONCEPTS

Dorsally situated iris tissue in a salamander eye can give rise to a lens, but only if a differentiated lens is not in the vicinity.

During this lens "regeneration," differentiated pigment cells lose original phenotype, undergo mitosis, and then differentiate in the new direction as lens fiber cells.

Many adult tissues may contain negative feedback inhibitors of mitosis called "chalones."

It is believed that chalones are distributed systemically and are specific for tissue type.

REFERENCES

Lens regeneration:
J.R. Ortiz *et al.* 1973. *Proc. Natl. Acad. Sci.*, 70, 2286. Cyclic AMP and cell-shape changes like those seen during the early phases of lens regeneration.
T. Yamada. 1967. *Current Topics Develop. Biol.*, 2, 247. The first place to go for references and general discussion up to 1967.
J. Dumont and T. Yamada. 1972. *Develop. Biol.*, 29, 385. More recent references and experiments on the iris to lens transition.

Neural retina regeneration from pigmented retina:

A.J. Coulombre and J.L. Coulombre. 1965. *Develop. Biol.*, *12*, 78. A study in chick embryos, paralleling earlier work in amphibians, that shows the most amazing transition of cells from the single layered pigment retina to the multilayered neural retina.

Chalones in general:

A.L. Thornley and E.B. Laurence. 1975. *Intern. J. Biochem.*, *6*, 313. A brief review on all the major chalones, and includes the key references on each of them.

Chalones, kidney and liver:

N.L. Bucher and R.N. Malt. 1971. *Regeneration of Kidney and Liver.* Little, Brown. A critical interpretation up to 1971 of the most studied cases of compensatory hypertrophy.

Chalones, other than kidney and liver:

J.C. Houck and A.M. Attallah. 1975. In F.F. Becker, ed., *Cancer*, vol. 3, p. 287. Plenum. A recent review of chalones emphasizing lymphocytes and other systems (not liver and kidney). Other articles in the different volumes of *Cancer* are pertinent to many of the topics in this book.

Chapter Thirteen:

A "glial" cell from a chick embryonic ciliary ganglion that shows intensive "ruffling" activity on much of its periphery. The folds and hair-like extensions seen here are dynamic structures that appear, move about, and disappear back into the cell surface. Thus, they are unlike cilia or sperm tails, which are long-lived organelles with complex internal architecture.

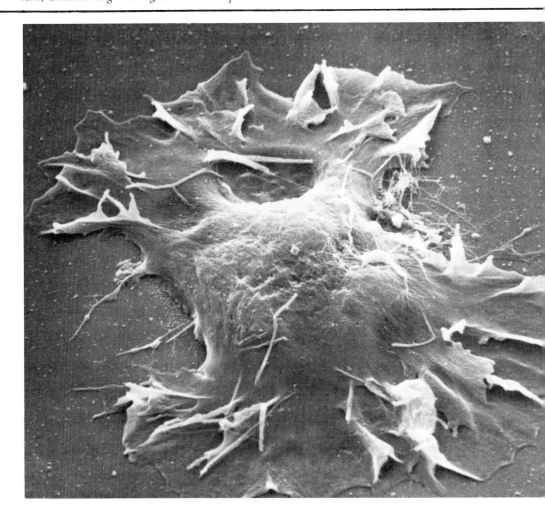

Cell Surfaces
and Development

The cell surface is the site where extrinsic factors act to affect cell movement, mitosis, and other developmental processes.

In Chapters 13, 14, and 15 we shall consider some of the possible mechanisms employed by cells and tissues as they interact with one another. First, we discuss the cell surface and its role in important developmental activities of cells. Then, we turn to evidence of cell-to-cell contact during embryonic tissue interactions. A brief discussion of molecular exchange between interacting tissues follows. And, finally, components of the extracellular spaces are described and evaluated in relation to cell contact and to complex phenomena such as morphogenesis. The purpose of these chapters is to provide further perspective on the ways that tissue interactions elicit morphogenesis and differentiation.

In the preceding chapters we have surveyed a variety of tissue interactions in embryos and adults; again and again, we have found that biological information originating from outside a cell influences restrictive or expressive aspects of cellular behavior. The cell surface, as an interface between extrinsic information and the cell genome, now deserves our attention.

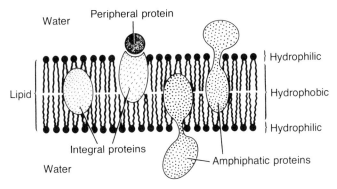

Figure 13.1

A schematic section through the plasma membrane. The lipid bilayer has hydrophilic faces and a hydrophobic interior. "Integral" proteins do not have substantial surface charge and can be found in the hydrophobic region. "Peripheral" proteins may be associated with portions of integral proteins that protrude into the aqueous phase (those portions would be of hydrophilic character, e.g., charged). "Amphiphatic" proteins each have two dissimilar parts, one soluble in lipid and the other in water. The protruding portions of amphiphatic proteins, or peripheral proteins, may have the ability to interact with environmental or cytoplasmic molecules. (After Singer.)

Any discussion of the cell surface must consider the latest revolutionary advance in our knowledge about cells: the dynamic nature of the plasma membrane. The plasma membrane can be imagined as a thin layer of oil that covers a tiny drop of water, the cytoplasm. In fact, the membrane consists of a double layer of lipid molecules, each of which has a hydrophilic (having affinity for water) and a hydrophobic (lacking affinity for water) end. Adjacent lipid molecules in the outer layer of the bilayer have their hydrophilic ends pointing outside the cell; lipid molecules in the inner layer of the bilayer have their hydrophilic ends pointing toward the cytoplasm. As is shown in Figure 13.1, the center of the bilayer is therefore composed of hydrophobic lipid chains. Such a region can contain only lipid-soluble, nonionic substances.

Proteins that are lipid-soluble and have no surface charge may reside in the hydrophobic region within the lipid bilayer. Such molecules have been termed "integral" proteins. Other proteins may be associated with the outer hydrophilic portions of the bilayer; they are called "peripheral" proteins. Finally, imagine a single molecule or molecular complex that has two discrete regions, one that acts like a peripheral protein, the other like an integral protein. The first part might protrude outside the cell or into its cytoplasm; the second, hydrophobic part would be embedded within the core of the bilayer.

Obviously, the protruding peripheral parts are of potential importance for interactions with extracellular agents or with cytoplasmic
organelles. Related to these interpretations are the facts that some
peripheral proteins are found *only* on the outer surface of the bilayer,
and others *only* on the inner surface—there is "asymmetry" in membrane structure.

Now, let us turn to the dynamic nature of the plasma membrane.
Protein molecules may be free to move about in the plane of the membrane. Imagine, for instance, that we insert a group of protein molecules into the lipid bilayer at one pole of a cell. If temperatures are
normal, those molecules will diffuse in the bilayer and become distributed over the cell surface. Singer and Nicolson have likened this
behavior to icebergs moving in a thin "sea" of lipid.

Movement of plasma membrane proteins in this "sea" may be
restricted in several ways. For instance, in lymphocytes of the immune
system, antigens (substances capable of eliciting antigenic response)
may bind initially at random over the cell surface (to peripheral receptors). During the following minutes, the bound antigen molecules
and the antigen receptors to which they adhere move over the cell
surface. First they form small patches, and then the patches move over
the surface to one end of the cell, thereby forming a "cap" of antigen
and receptor. Various of the cytoplasmic filament systems (microtubules, microfilaments) are involved in these directional displace-
ments of cell surface components. However, we do not know yet
precisely how such cortically located organelles interact with the receptors, part of which are presumably immersed in the hydrophobic
internal region of the lipid bilayer. Nevertheless, this phenomenon
of interaction between an external molecule (the antigen), membrane
protein (the receptor), and cytoplasmic organelles gives us a model of
great significance for basic activities of developing cells (locomotion,
cell-to-cell communication, etc.).

Another means of affecting the distribution of molecules in the
cell surface is provided by cell-to-cell junctions. As seen in Kelly's
diagram (Figure 13.2), there are many types of junctions that can form
between cells. Epithelial cells in particular have a variety of junctions
on their lateral surfaces near both ends of the cell. One of the many
functions of such junctions is to limit the freedom of diffusion of molecules that are immersed in the lipid bilayer. As a result, separate "do-
mains" of molecules called intramembranous particles may exist (see
Chapter 16 and Figure 16.4). Thus, some substances may be restricted
to the ends of cells, beyond the encircling junctional complexes, others
perhaps to the lateral surfaces, and so on. Now, imagine for a moment

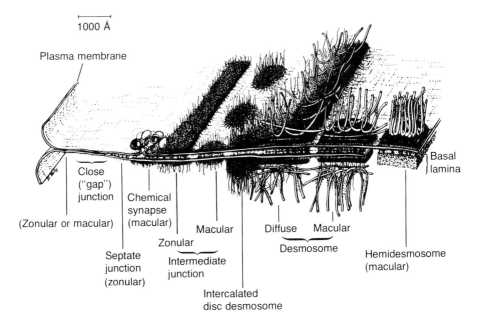

1000 Å

Plasma membrane

Basal lamina

Close ("gap") junction
(Zonular or macular)

Chemical synapse
(macular)

Macular

Zonular

Septate junction
(zonular)

Intermediate junction

Intercalated disc desmosome

Diffuse

Macular

Desmosome

Hemidesmosome
(macular)

Figure 13.2

Kelly's summary of the types of junctions between cells of higher organisms. To the left and nearest the end of these cells are found the occluding and gap junctions. In the first of these the outer layers of the plasma membranes appear to fuse and the junction acts to prevent passage of most foreign substances to the space between the lateral surfaces of cells. The gap junction is the probable site of electrical continuity between two cells and the point where small molecules may be exchanged (see Chapter 16). The intermediate junction region (center) is a place of firm adhesion between cells, and the point where actin filaments are associated with the inner surface of the cell surface. Proceeding further to the right in the diagram, the region of desmosomes is encountered. Here, too, intercellular adhesion is strong; and, within the cytoplasm long "tonofilaments" are seen to loop into the substance on the cytoplasmic side of the plasma membrane. Some believe that these tonofilaments are skeletal agents. Finally, at the right, the situation is diagrammed in which a half "hemi-"desmosome is shown opposite a portion of basal lamina material (such an arrangement would only be seen at the basal surface of epithelial cells, but is shown in this position for convenience). Note that the intermediate and desmosomal junctional complexes can occur either as continuous bands ("zonular") or as discrete patches ("macular"). (Courtesy of Douglas E. Kelly.)

that such substances or particles influence the ability of an epithelial cell to respond to a mesenchymal cell during a tissue interaction; obviously, the restriction of the particles to one domain as opposed to another could be an important limitation on the interaction.

Related to these considerations is the response of intramembranous particles to extracellular factors. If, for instance, the pH of the fluid that is bathing cells is lowered, then the particles rapidly aggregate within the membrane; raising the pH again may result in particle dispersal. Alternatively, contact or close proximity between the surfaces of two cells may affect particle distributions or the placement of molecules that can bind various foreign substances. Finally, exposure of sensitive cells to the peptide hormone, vasopressin (a substance that causes osmotic water flow across epithelia, such as in the toad's bladder), results in remarkable aggregation of intramembranous particles just on the cell surfaces where water movement occurs. Here, where a foreign protein is acting, and in the other sorts of cases outlined above, we have models for the way that cells could respond to contact with other cells or with elements of the extracellular environment (as in Chapters 14 and 15). The important conclusion is, therefore, that a variety of external or internal factors can influence the distribution of molecules in the cell surface. We shall return to this point after considering aspects of cell locomotion and mitosis in relation to the cell surface.

Contact Inhibition Phenomena

Both cell locomotion and mitosis can be influenced by the cell surface. When two normal fibroblasts that are moving over the bottom of a culture dish meet and touch one another, their forward advance is halted. Michael Abercrombie and his collaborators (Ciba Symposium, 1973) have shown that this "contact inhibition" results in rapid paralysis of the anterior locomotory region, the ruffled membrane (see Figure 13.3). New ruffles then appear elsewhere on the periphery of the cell. Soon, one new ruffling area predominates and leads the cell off in a different direction. If cell-population density is raised so that such encounters are frequent, a given cell shows much less net movement. The relative immobilization resulting from this contact inhibition phenomenon may contribute substantially to the stability of tissues.

Figure 13.3
A sketch of two cells displaying contact inhibition of directional locomotion. As cell A moves forward, a portion of its anterior surface comes in contact with cell B. That portion of the anterior surface is "paralyzed": ruffling and further movements outward cease. Extensive ruffling appears elsewhere on the surface of cell a, and it moves off in a new direction. (See J.P. Trinkaus et al., Exptl. Cell Res., *64 (1971), 291.)*

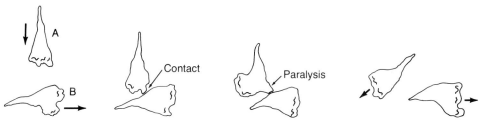

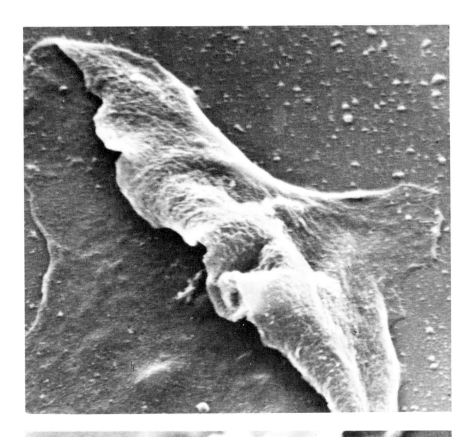
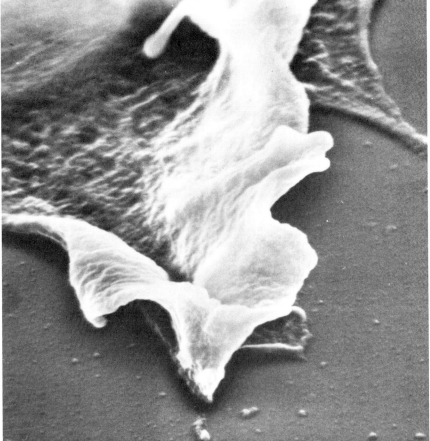

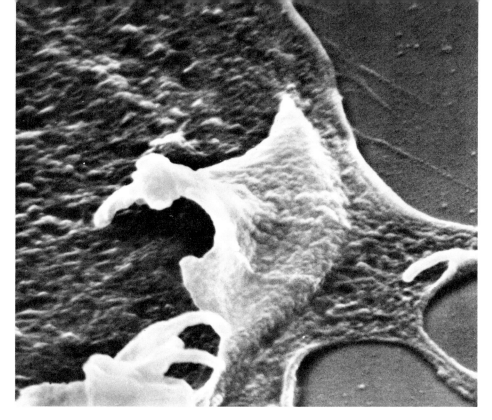

Figure 13.4
*"Ruffles" on glial cells of embryonic ciliary and sensory ganglia. **A, B, C**: Note the wave-like shapes of these dynamic extensions of the cell surface. The ruffles tend to appear at the margin of the cell, extend upward into the fluid medium, and then fold back onto the upper surface of the cell as they disappear. In **D**, large ruffles appear on the glial cell as it encounters and encircles a nerve axon. Contact between these two types of cell surface does not lead to localized paralysis of ruffling and to contact inhibition of directional locomotion (see text).*

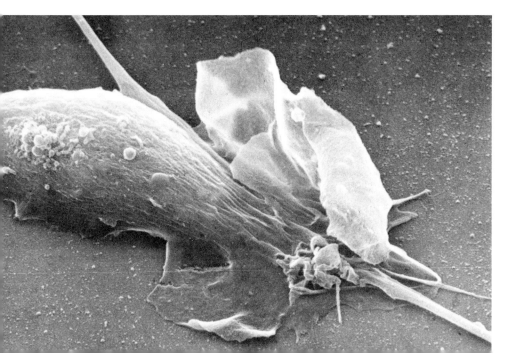

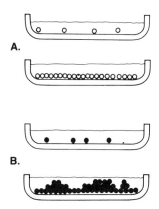

188

Figure 13.5

*Contact inhibition of mitosis. Normal cells (**A**) divide until they cover the bottom of the culture dish, forming a monolayer. Then, their division rate falls precipitously. In contrast, some cancer cells or virus-transformed cells (**B**) go beyond the monolayer condition and pile up into multilayers because they apparently lack the contact-mediated inhibition of mitosis. (After J.D. Watson,* The Molecular Biology of the Gene, *W.A. Benjamin, 1976.)*

Biologists working on quite different phenomena observe another kind of "contact inhibition" when they allow normal cell lines to undergo substantial mitosis in a culture dish. As the bottom of the dish becomes entirely covered with a single "monolayer" of cells, mitosis slows and halts (see Figure 13.5). Contact with sufficient numbers of neighbors seems to cause this inhibition, not (for a number of cell types) depletion of nutrients or equivalent uninteresting possibilities.

Virus transformation of cells or certain cancerous conditions add considerable perspective to the contact inhibition story. Certain types of cells can be exposed to a virus and caused to undergo a "transformation" process. From our point of view, the interesting consequence of this change is that transformed cells do not halt when they meet; they simply crawl right over or under each other! Furthermore, mitosis does not stop at the monolayer stage; division continues until many cell layers pile up and nutrients do indeed become limiting. Both types of contact inhibition are lost in certain cancerous conditions characterized by rapid mitosis and by invasiveness of foreign tissues.

The surface of transformed cells actually shows altered antigenic markers, that is, a change in the type of molecules on the surface that can elicit antibody production. Normal cells apparently "recognize" the surface of other normal cells (and show the contact inhibition syndrome) but cannot recognize transformed cells or some cancer cells—they crawl under or over the transformed one and simply ignore its presence.

How does contact inhibition operate? Stabilization of the cell surface at the points where cells meet (i.e., immobilization of membrane movements or flow) could account for the phenomenon. Alternatively, altered function or distribution of cytoplasmic microtubules or microfilaments (see Chapter 7) may be related to the inhibition of cell movement in a particular direction. A good possibility is that the concentration or distribution of calcium ions and cyclic AMP might be affected by contact with another cell. To understand some of these alternatives, we must pause and consider cyclic nucleotide levels as a function of surface-active hormones.

Hormones and Cyclic Nucleotides

We have already learned that the pancreatic mesenchymal factor that stimulates mitosis can act at the cell surface even when the factor is bound to Sepharose beads. A class of small proteinaceous hormones, including insulin, nerve growth factor, epidermal growth factor, and somatomedin (a substance released from liver in response to the pituitary's growth hormone and which causes mitosis and growth in bone and other tissues) apparently behave in the same manner. Such hormones bind to receptor molecules that are located at the cell surface and then exert a primary physiological effect. Perhaps because of similarity in structure, insulin, somatomedin, and epidermal growth factor compete for binding to the same receptor. On the other hand, a diversity of receptor types may be present on a single differentiated cell. Eight different hormones bind to eight distinctive receptor types on fat cells, yet all elicit the same primary response: inactivation of the cell surface enzyme adenyl cyclase (the enzyme that manufactures cyclic AMP). This causes the amount of cyclic AMP to decrease and a variety of secondary responses.

Since insulin, somatomedin, and, perhaps, other proteinaceous hormones act similarly, we can use these illustrations to construct a simple model pertinent to tissue interactions. Cuatrecasas has proposed a "mobile receptor hypothesis" wherein the binding of a hormone to a receptor molecule permits the complex to move in the plasma membrane and to interact with adenyl cyclase. The association may

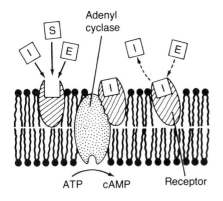

Figure 13.6
A diagrammatic view of the plasma membrane of an animal cell. The receptor can bind insulin (I), somatomedin (S), or epidermal growth factor (E). When one of the hormones is bound, the receptor-hormone complex interacts with adenyl cyclase to change that enzyme's activity. "Competition" in binding to the receptor could occur if more than one hormone is present (right). (After Cuatrecasas.)

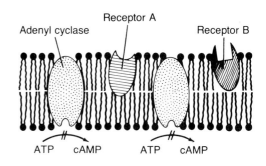

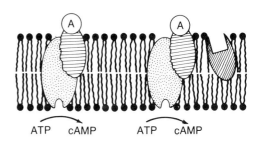

Figure 13.7
The "mobile receptor" hypothesis. When hormone A binds to its receptor, the complex moves in the plane of the plasma membrane's lipid bilayer until it encounters and interacts with an adenyl cyclase molecule. That enzyme is activated, amounts of cyclic AMP increase, and cell responses are seen (say, inhibition of cell division cycles). The hypothesis is particularly attractive for cells that have many different types of receptors, all of which can interact with adenyl cyclase. In such cells, the same enzyme molecules could be used by the many different activating hormones. (After Cuatrecasas.)

alter that enzyme's activity, either by simply inhibiting it or by altering its substrate specificity. Consequently, less cyclic AMP is produced, and in some cells *more* cyclic GMP is manufactured.

What is the significance of changes in amounts of cyclic nucleotides? Cultured cells that are undergoing rapid mitosis and are moving freely about dishes have little cyclic AMP and much cyclic GMP (see Figure 13.8). As cell density increases and contact inhibition phenomena begin to manifest, cyclic AMP increases, cyclic GMP decreases. Virus-transformed cells or some cancerous cells (most of which show an absence of contact inhibition) have little cyclic AMP even when they are cultured at high densities. It is as if the link of cell surface to adenyl cyclase regulation is absent or deficient.

Measurements of amounts of cyclic nucleotide in normal tissues, including ones stimulated by the hormones, reveal the same general relationship; that is, the amount of cyclic AMP is inversely correlated with DNA synthesis (and mitosis). One might guess, therefore, that regulatory pathways for mitosis could operate by affecting the enzymes that manufacture or hydrolyze that nucleotide. In fact, work with the toxin molecule produced by the cholera bacterium provides an illustration of how a chalone might operate via these mechanisms. Cholera toxin binds to a ganglioside (a glycolipid) in the cell surface; the complex then interacts with adenyl cyclase, and the amount of cyclic AMP increases dramatically. As expected, DNA synthesis decreases and mitotic cycles halt. In formal terms, the cholera toxin acts just as a chalone might in inhibiting mitosis.

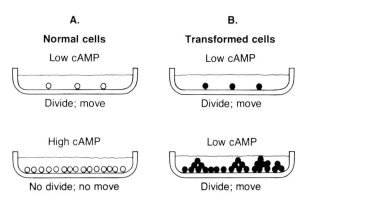

A.
Normal cells

Low cAMP

Divide; move

High cAMP

No divide; no move

B.
Transformed cells

Low cAMP

Divide; move

Low cAMP

Divide; move

Figure 13.8
*Cyclic AMP in relation to contact inhibition of mitosis and directional locomotion. As the number and density of normal cells increase (**A**), the contact inhibition phenomena operate and the amount of cAMP increases. In contrast, when density of transformed cells increases, contact inhibition is not seen and the amount of cAMP remains low. (Note: by "move" or "no move" is meant directional locomotion; contact inhibition does not necessarily inhibit locomotion itself.)*

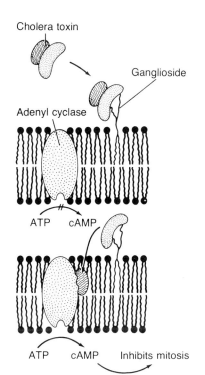

Cholera toxin

Ganglioside

Adenyl cyclase

ATP   cAMP

ATP   cAMP   Inhibits mitosis

Figure 13.9
A model for chalone action. Cholera toxin has two subunits, one of which is lipid soluble but cannot act on adenyl cyclase by itself. Cuatrecasas hypothesizes that the toxin interacts with the ganglioside in the plasma membrane, and that the lipid-soluble subunit of the toxin can then interact with adenyl cyclase. That enzyme is activated, cAMP increases, and mitosis decreases. If products of differentiated cells in a tissue acted similarly on the adenyl cyclase of germinal cells in the same tissue, then a chalone system of mitotic inhibition would be operating (see Chapter 12). (After Cuatrecasas.)

Figure 13.10

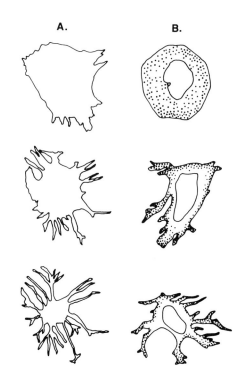

*Cyclic AMP and lens regeneration. **A.** Tracings of a cell removed from the dorsal iris of a salamander and subjected to increased amounts of cAMP in its cytoplasm. The cell assumes a highly "stellate" configuration, which is remarkably similar to behavior of the pigmented dorsal iris cells during the early parts of lens regeneration. In **B,** sketches of pigment cells in the intact iris are shown at stages following lens removal. Many pigmented cells form long extensions filled with pigment granules; such extensions are broken off or removed by macrophages (cells that "chew up" foreign materials). This is an important part of the dedifferentiation phase, before mitosis and generation of the new lens cell population. The observations with cAMP in culture suggest that the same mechanism might operate during the early phase of the iris-to-lens transition. (After J.R. Ortiz, et al., Proc. Natl. Acad. Sci., 70 (1973), 2286. See also Develop. Biol., 29 (1972), 385.)*

It is obviously tempting to explain contact inhibition of mitosis and of directional locomotion in this framework. Thus, increased cell density leads to frequent contact between cells, the equivalent of hormone-receptor interactions occur, nucleotide amounts shift, and mitosis halts. It would be a mistake, however, to extrapolate too far from work on adult cells. First, the basic hypothesis is still based on circumstantial evidence. In addition, recall that periodate-inactivated mesenchymal factor only stimulated pancreas cell mitosis when amounts of cyclic AMP levels are made *high*. Or consider that some protein hormones do not seem to act via cyclic AMP; nerve growth factor and erythropoietin, the kidney hormone that stimulates red blood cell production, are examples. Furthermore, some developmentally important hormones—epidermal growth factor is one—take long times (about 6 to 8 hours for EGF) to evoke a response, though the changes in cyclic nucleotides that we have been discussing occur in seconds or minutes. Finally, recall that the greatest amounts of cyclic AMP in salamander limb regenerates are present just when mitotic activity is greatest too. Obviously, interpretation is not easy when measurements must be performed on complex tissues that contain many cell types, some of which may be dividing and others not. Nevertheless, the simple extrapolation from cells in culture to cells in tissues and organs and embryos may be dangerously misleading.

Clearly, cyclic AMP will not solve all our problems. What is important, therefore, is the concept that cell-to-cell or foreign molecule-to-cell interactions can act at the cell surface to initiate a chain of events with important developmental consequences.

Significance of Contact Inhibition for Tissue Interactions

The work on cell surfaces and the contact inhibition phenomena has implications for both mature and developing tissues. First, it relates to the stability of the differentiated phenotype. Under normal circumstances in adult functional tissues, cells are immobile and divide at low rates that are coupled to the rate of cell death. However, most such cells have latent capacities to move and to divide rapidly, but do not do so because of the stability of the local tissue environment. Contact inhibition tells us that one ingredient of the control circuit may be the cell surface and its capacity to respond to proximity of neighboring cells. Wounding, removal of inhibitors, removal of nerve, or other perturbations of a cell's environment, all might operate by triggering events at the cell surface that elicit the basic cellular responses of mitosis or locomotion. If this argument is correct, then an important part of cell differentiation must be the construction of cell surfaces that are capable of participating in the control circuit. Receptors, transducers (such as adenyl cyclase), and surface-associated effectors (microtubules?), would be just as important ingredients of the final differentiated state as are the primary differentiative characteristics of a cell line (sarcomeres for muscle; secretory apparatus for pancreas; etc.).

A second implication of cell surfaces and contact inhibition relates to the early phases of development. Adhesive and antigenic properties of early embryonic cells change in time. This may be an important component of morphogenesis, both for single cell locomotion and for cell group movements (as discussed for salivary morphogenesis, Chapters 7 and 15). A good illustration of the significance of changing cell-surface properties is seen in the limb mutant, *talpid*³, of domestic chickens.

Donald Ede, an English biologist, has found evidence that limb mesoderm cells from *talpid*³ embryos are less adhesive to one another than are cells from normal embryos. The mutant cells also accomplish

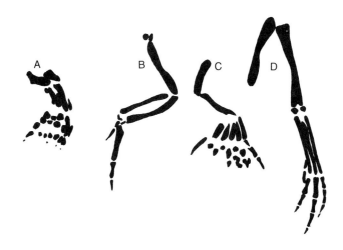

Figure 13.11
*Tracings of photographs of limb bones from talpid² and normal embryos: **A.** talpid² wing;*
***B.** normal wing; **C.** talpid² leg; **D.** normal leg (all from 9.5-day-old embryos). Note the*
marked shortening and broadening of the mutant limbs. The polydactylous (extra digit) nature
of the mutation is obvious in both wing and leg. These talpid² limbs are generally similar to
the talpid³ ones used for experiments described in the text. The short, broad mutant limbs may
arise because of deficiencies in adhesive and locomotory properties of the limb mesoderm cells.
(From P.F. Goetinck and U.K. Abbott. J. Exptl. Zool. 155 (1964), 162.)

less net cell movement than normal cells. If individual cells are followed carefully, it proves that mutant cells move quite normally when
they are moving; what the mutation does is to cause *talpid³* cells to
spend abnormally long intervals in an immobilized condition. Obviously, we would like to know whether periodic fluctuations in enzyme activities that lead to cyclic nucleotide synthesis or degradation
could account for the intermittent locomotion. What is so intriguing
about the *talpid³* mutation is that the apparent alteration in cell adhesivity and locomotory frequency can easily account for the gross
anatomical effect of the mutation. *Talpid³* limb buds are broad and fan
shaped, and give rise to limbs that are unusually wide and polydactylous (see Figure 13.11). Computer simulations of limb development
suggest that just such a morphology would be generated if limb mesoderm cells have decreased motility. Thus the observed properties of
individual mesoderm cells from the mutant embryos are precisely what
is required to produce the gross morphological effect.

Speculations about the Cell Surface and the Restricted Condition

These kinds of observations mean that cell surfaces may change in developmentally significant ways during the early expressive phase of cell development, i.e., *before* the restriction process is complete. Is it possible that the cell surface is involved in the restriction phenomenon itself? Does the surface act as an intermediary for filtering and transmitting signals originating outside the cell?

The examples of the gray crescent and the pole plasm established that the cytoplasm and perhaps the cell cortex can influence which genes a line of cells can ultimately use. Do properties of cell surface, originally derived from gray crescent materials, act so that only signals eliciting mesodermal family restriction and expression succeed in reaching the genetic apparatus of a prospective mesoderm cell, whereas signals for ectodermal and endodermal responses are rejected at the cell surface?

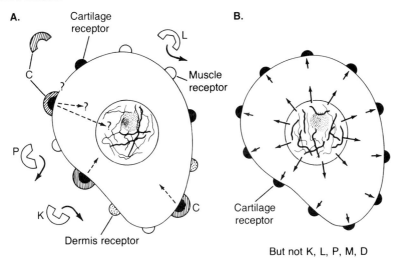

Figure 13.12

In this hypothetical scheme, surface factors (receptors?) derived from the gray crescent control whether or not foreign information for cartilage (C), kidney (K), lens (L), or pancreas (P) development can exert an effect on this amphibian mesoderm cell. Whether such information acts at the surface, in the cytoplasm, or in the nucleus is unclear. Once the cell has become determined (B), its chromatin has the capacity to show cell-specific transcriptional properties. The mRNA types that reach the cytoplasm and are translated into protein govern the types of cytoplasmic and cell surface molecules that will be present, including receptors for further extrinsic regulation. Thus a kind of feedback stability may be set up, in which nucleus and cytoplasm could continuously contribute to stability of the determined state.

196   We cannot as yet answer such questions. Nevertheless, in each of the next three chapters we will encounter cases in which contact of the cell surface with foreign molecules, or an hypothesized release of developmental information from the cell cortex to the inside of the cell, elicits gene function and production of specific proteins. Until proven otherwise, therefore, the cell surface may be considered a prime candidate for being a site for factors that control both restrictive and expressive events in animal development.

CONCEPTS

Contact between cells can affect the direction of cell locomotion or the ability to carry out mitosis.

Receptors for many protein regulatory molecules are located on the cell surface.

A number of protein regulatory molecules seem to exert their effect at the cell surface by changing the activity pattern of adenyl cyclase, the enzyme responsible for manufacture of cyclic AMP.

Because of its composition and properties, the cell surface may play an important role during the restriction and determination processes, and also later, as a contributing factor to the heritable stability of the determined state.

REFERENCES

Cell locomotion and contact inhibition:

Locomotion of Tissue Cells. Ciba Foundation Symp. 14 (new series), 1973. Elsevier. A wealth of observations on contact inhibition and locomotory properties of many cell types, including studies by Abercrombie and his group.

A. Harris. 1974. In R. Cox, ed., *Cell Communication.* Wiley. A superb critical review of contact inhibition in relation to locomotion, mitosis, and cancer.

Cell membranes and surfaces: 197

S.J. Singer and G.L. Nicolson. 1972. *Science*, *175*, 720. This paper has had greater impact than any other in recent years in shaping research and ideas on the cell surface.

K.F. Heumann. 1973. *Fed. Proc. Amer. Soc. Exptl. Biol.* 32, 19. A series of research papers on widely divergent aspects of cell surfaces that introduces the reader to several aspects of cell surfaces in development not covered in this book.

S.J. Singer. 1976. In *Surface Membrane Receptors*. Ed. R. Bradshaw. Plenum. Recent literature and speculations on the fluid, mosaic model of membrane structure.

Hormones and cell surface receptors:

P. Cuatrecasas *et al.* 1975. *Rec. Prog. Hormone Res.*, *31*, 37. A review of hormone receptor interactions and how they may affect membrane functions and, in turn, cell behavior.

Growth factors:

R.O. Greep, ed. 1974. *Rec. Prog. Hormone Res.*, *30*. This volume contains a series of articles on EGF, NGF, insulin, and somatomedin.

***Talpid*[3] mutants:**

D. Ede. 1975. *J. Cell Sci.*, *18*, 301. This issue contains papers concerning the *talpid* mutants, cell locomotion, and limb morphology. References to the computer modeling are included.

Cell surfaces in development, in general:

A.A. Moscona, ed. 1974. *The Cell Surface in Development*. Wiley. Cell recognition, low resistance coupling, GAG, nerve cell development, and many other intriguing topics are found in these 16 papers.

Cyclic AMP during limb regeneration:

J. Jabaily *et al.* 1975. *J. Morphol.*, *147*, 379. This paper includes a useful summary of work on cyclic AMP in complex systems.

Chapter Fourteen:

A section through a 22-micron-thick filter, with mouse embryonic kidney mesenchyme cells above and spinal cord cells below. The pores of the filter twist back and forth in space so that, in this ultrathin section, only short segments of cellular cytoplasmic extensions (C) are seen. In fact, if we could view such processes in three dimensions, they would be seen to be connected with cells above or below the filter. Kidney tubules will form in the tissue above the filter in response to the contact between the extensions of the two cell types. (Courtesy of E. Lehtonen and L. Saxén. From J. Embryol. Exptl. Morphol., *33 (1975), 187.)*

Cell Contact and Tissue Interactions

Cells can "contact" one another though separated by seemingly long distances. What does "contact" mean?

When tissue interactions occur, do cells of the two tissues actually touch each other? Or are extracellular molecules the vehicles of interaction? We have already discussed some cases in which important effects can be mediated by a fluid environment (recall the AER maintenance factor effect). In this chapter we will examine these questions directly.

"Contacts" between Interacting Tissues

The adult, metanephric kidney of birds and mammals includes a branched, tree-like collecting duct system, and a series of functional units, the nephrons. Both parts arise from mesodermal cells. The duct portion develops from the ureteric bud by a type of morphogenesis similar to that which occurs in salivary glands or lungs.

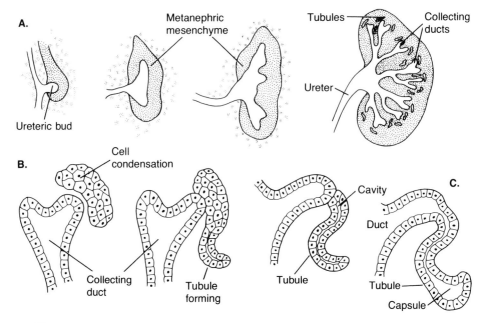

Figure 14.1

*Mammalian kidney development. **A**. The ureteric bud branches within the investing meta-nephric mesenchyme and gives rise to the collecting duct (pelvic) portion of the metanephric kidney. At many places near the tips of the branching ureteric bud system, focal condensations of mesenchyme cells appear (**B**), gradually form an epithelium with a central cavity, and ultimately link up (**C**) with nearby portions of the ureteric collecting duct system. The individual tubules subsequently develop into specialized regions of the adult kidney (Bowman's capsules; proximal and distal convoluted tubules; loop of Henle; etc.).*

Near the tips of the branching duct system, the populations of adjacent mesoderm cells interact with one another. As a result, focal condensations of mesoderm appear. Cells in the condensations form an epithelium, which folds and twists to generate "S"-shaped tubules, the precursors of the nephron parts.

Grobstein, Borghese, and others established that the ureteric bud does not branch if it is not in the vicinity of the surrounding kidney mesoderm (termed metanephrogenic mesenchyme). Conversely, the little tubular systems do not form if the ureteric bud does not act on the nephrogenic mesenchyme. This is an excellent example of reciprocity in interactions; neither tissue component develops normally in the absence of the other.

The capacity to stimulate kidney tubule development is not limited to the ureteric bud. Pieces of the dorsal spinal cord and several types of embryonic epithelia or mesenchymes can elicit the process in nephrogenic mesenchyme.

Saxén and his associates have established that proximity of inter- 201
acting cells is required to stimulate condensation formation and tubule
development. Previously, it had been observed that if nephrogenic
mesenchyme and spinal cord were placed on opposite sides of a porous
filter, kidney tubules would form (Figure 14.2). Since electron microscopy did not reveal cellular extensions that crossed the tiny pores
(diameters ca. 0.4 microns) of the filter, it was concluded that cell-to-
cell contact was not required during the interaction. Using new fixa-
tion and microscopic methods, Saxén's group has now shown that
long, thin cellular extensions can penetrate the filters (see the frontispiece to Chapter 14). Furthermore, kidney tubules can form only if
this penetration takes place and only if the interacting cell surfaces
come very close to each other. Finally, these researchers have found
that the time necessary for the cellular extensions to penetrate the filter
corresponds to the length of time the spinal cord must be present if
kidney tubules are to form. That is, no tubules form in the mesenchyme if the spinal cord is removed from the filter after 12 hours (the
extensions have not penetrated deeply then); they do form if the spinal
cord is scraped off at 24 hours (when penetration is advanced).

These observations show that the factors which initiate condensation formation and tubule morphogenesis cannot act effectively
through the fluid-filled spaces of a filter. But does direct cell-to-cell
contact ever occur in an embryo? Electron microscopic examination of
salivary glands, lungs, and teeth suggest that it does (see Figures 14.3

Figure 14.2
*An experimental culture in which nephrogenic mesenchyme is above the porous filter and
spinal cord tissue is below. Tubules, seen here in section, have formed in the mesenchyme in
response to the action of spinal cord (which mimics ureteric bud in this respect). (Courtesy of
L. Saxén.)*

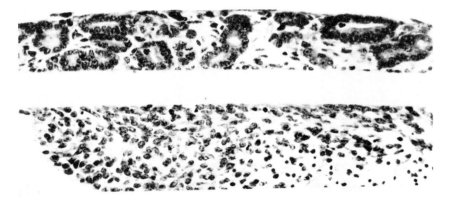

202   and 14.4). We shall see in Chapter 15 that the basal surface of epithelial cells is in contact with a sheet-like basal lamina, which is a complex of collagen and other macromolecules. It has been assumed that this layer of extracellular materials is an effective barrier against direct contact between epithelial and mesenchymal cells. However, recent observations on the organs listed above reveals that the basal lamina may be absent or degraded in specific places and at specific times during development (see Figure 14.4). Even where a seemingly intact basal

Figure 14.3
The interface of mesenchyme and epithelium in a developing tooth germ. A mesenchymal cell extension (E) is seen protruding through the dense mat of collagen fibers (C). The extension appears to penetrate the basal lamina (L), which has undergone some degradation by this stage of tooth development. The extension then passes between two epithelial cells. This is a representative case of cell-to-cell "contact" in embryos. (Courtesy of H.C. Slavkin and P. Bringas, Jr. From Develop. Biol., *50 (1976), 428.)*

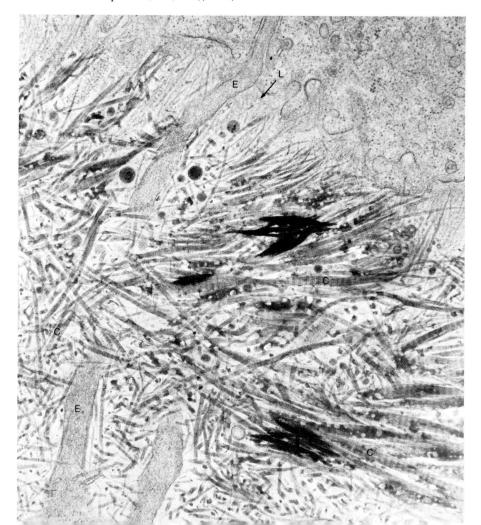

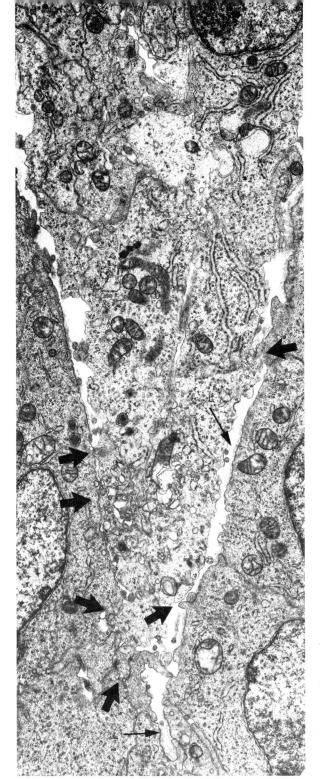

Figure 14.4
*The "cleft" at the surface of a mouse
salivary gland epithelium (as in Figure
7.6). Epithelial cells appear to the lower
left and right, and portions of mesenchyme
cells fill the centrally placed cleft. Note
the fragmentary nature of the basal
lamina (L) and the numerous sites of
direct "contact" (arrows) between the two
cell types. This shows that long cellular
extensions are not the only means by
which interacting cells touch one another.*

204   lamina appears to be present, diligent searching with the electron microscope may reveal narrow cylindrical cytoplasmic extensions that penetrate the lamina and so permit "contact" of epithelial and mesenchymal cell surfaces.

Perhaps the most surprising feature of the interface between interacting tissues comes from viewing various surfaces with the scanning electron microscope (Figure 14.5). Though mostly covered by basal lamina (Figure 14.5 A), there are in fact large numbers of long hair-like extensions protruding from the basal surface of epithelial cells that are engaged in tissue interaction (Figures 14.5 B and C). Such cellular extensions provide a greatly expanded surface area that could be significant during interactions. Cytoplasmic extensions of varying length and number also originate from mesenchymal cells (Figure 14.5 D), and twist through the extracellular spaces. Whether contact between the epithelial and mesenchymal cell processes is the actual mediator of tissue interactions is not known yet, although that certainly is an attractive hypothesis.

Saxén's experiments on the kidney establish that cell-to-cell contact is required. But what does "contact" mean? The surfaces of most vertebrate cells are believed to be coated with complex mixtures of proteins and sugars. As far as is now known, these materials may be present at the points of contact between spinal cord and kidney mesenchyme cells. Furthermore, no special "gap" junctions (Figures 13.2, 16.3, and 16.4) are known to be present. We discuss functional properties of gap junctions later in this book, since ions and small molecules can be exchanged between cells through these junctions.

Though gap junctions have not as yet been seen where cell extensions "touch," and though the cell coating of proteins and sugars is apparently present at these places, the effects of the proximity between cell extensions could be mediated by the plasma membrane. Recall that extracellular conditions (e.g., pH), or proximity to other cells or to a solid substratum, can affect the distribution of particles in the plasma membrane. Changes in distributions of intramembranous particles or of integral or peripheral proteins could in turn alter intracellular structures or processes in a way that propagates a developmental effect. Therefore, contact does not necessarily imply the necessity to exchange informational molecules.

Conclusions about the kidney system do not apply to all tissue interactions. For instance, when mesoderm and prospective medullary plate of amphibian embryos interact, cellular extensions do not have to penetrate filters; the extracellular environment is able to mediate the interaction. Here, in the pancreas, in the epidermis, and perhaps

Figure 14.5

A. A view looking directly at the surface of the basal lamina on the epithelium of a mouse embryonic lung bronchus (as on one of the tips in the frontispiece to Chapter 7). Mesenchyme cells and collagen (see the frontispiece to Chapter 15) have been removed with enzymes to permit observation of the lamina. A few bulges and the outlines of epithelial cell extensions are seen in this photograph. B. A lung bronchus fractured so that the original mesenchyme side is seen at the top, and the original lumenal (cavity) side at the bottom. The basal lamina has been removed with enzymes so that hundreds of hair-like extensions can be seen on the upper surface (mesenchymal side) of this epithelium. C. The same specimen as in B, looking down at the edge of the fracture (as seen at the top of B). The numerous cellular extensions provide a huge surface area for potential interaction with mesenchymal cell extensions. D. An embryonic mouse lung showing the thick mat of collagen fibers (C) lying above the base of a bronchus. Mesenchyme cells are seen to the lower left, and from them exceedingly long cellular extensions (E) extend over and down into the collagen. (Courtesy of E.H-L. Hu.)

206   in other systems, a fluid-filled space can permit transfer of the required developmental information. This does not necessarily mean that the vehicles which transfer the information are in solution. Nor does it mean that the same information is transmitted over equivalently long distances (ca. 25 microns) in the embryo, since, in general, cells of interacting populations are much closer together than they are when separated by the 25-micron thickness of a filter.

An intriguing aspect of the "contact" relationship has been elucidated by Meier and Hay in developing cornea. In the next chapter we will examine the interaction between lens and head ectoderm which results in cornea formation. What is significant here are situations in which corneal epithelium and lens "capsule," an extracellular macromolecular complex, are placed on opposite sides of porous filters. Corneal cell extensions penetrate the filter and contact the capsule. The corneal cells then respond by increasing the rate at which they synthesize and secrete their specific products; e.g., macromolecules of the corneal stroma (thus, the cells are showing a differentiative response).

By varying the pore sizes and thicknesses of the filters, we find a linear relationship between: (1) the surface area of contact of the cell extensions and the lens capsule; and (2) the rates of synthesis. The greater the penetration and contact, the greater the synthetic response. Not only does this quantitative relationship operate in developing cornea, but a qualitative one may operate also. Thus collagen (one of the macromolecular types of the corneal stroma, to be described in the next chapter) is synthesized in response to cellular contact with collagen itself placed across the filter. More important here, is the observation that this matrix collagen does not have to enter the responding corneal epithelial cells to exert its stimulatory effect—it can act at the cell surface! We should certainly consider the possibility, therefore, that some tissue interactions operate by means of information transfer at the cell surface. Protein hormone-receptor interaction, mesenchymal factor of pancreas working though bound to Sepharose beads, contact-mediated redistributions in plasma membrane proteins, and many other examples parallel this finding that extracellular collagen can stimulate intracellular collagen synthesis. Moreover, recall the case of the insect hormone ecdysone which apparently affects membrane permeability so that Na^+ levels fall and K^+ levels rise in cells. As a result, puffing patterns of genetic loci change. Could contact between two cells across a filter, or of a cell with a substance such as collagen, elicit analogous changes in ion content and so trigger gene activity? We do not know. But this and the other examples cited bring

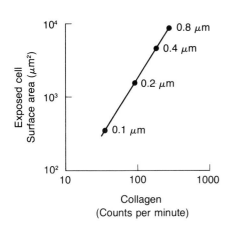

Figure 14.6

The relationship between the amount of newly synthesized collagen and the surface area of cell extensions protruding through a filter. This is a log-log plot showing that, as more of the corneal cell extensions are found on the lower surface of a porous filter (where they contact the stimulating tissue, killed lens capsule), there is a proportional increase in the quantity of radioactively labeled collagen in the culture. The four points correspond to the pore sizes of the filters that were employed in the experiments. (Redrawn from Meier and Hay.)

us again and again to the cell surface as a potentially important factor in cell and tissue interactions. It seems safe to assume, therefore, that the plasma membrane and its component molecules are destined to take a preeminent place in any future general theory of information transfer and processing in embryos.

Transfer of Macromolecular RNA?

On the last pages of this chapter, we will treat an immense subject with a few words. Saxén and Toivonen have already summarized much of the experimentation that has been designed to identify the instructional molecules that are responsible for development of the vertebrate central nervous system. Much of that work employs the strategy of applying tissue extracts to prospective nervous system tissue, of observing results, and then of analyzing the extracts for the responsible agents. Alternatively, possible transfer from one tissue to another of proteins or RNAs has been investigated, though usually without a clear concept of how such substances might act on responding cells.

What is the status of such experimentation in developing organs such as kidney, salivary gland, or lung? We shall see in Chapter 15 that one type of protein (collagen) made by mesenchyme can apparently pass through filters and accumulate near the surface of salivary epithelial cells. In this chapter we shall consider possible transfer of macromolecular RNA during such a tissue interaction.

Many technical difficulties are encountered in investigating such a question. For instance, if a precursor of RNA, say ³H-uridine, is

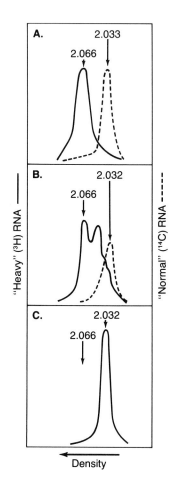

Figure 14.7

*Results of heavy-isotope labeling to trace RNA transfer between lung tissues. These three plots show the density profiles of extracted RNA. **A**. The peak on the left is the "heavy," density labeled RNA, that is extracted from lung mesenchyme after it has been exposed to the precursors for 24 hours. To the right in **A** is seen the density of normal lung RNA. **B**. RNA extracted from the mesenchyme after an additional 24 hours of culture, during which time the mesenchyme was interacting with unlabeled lung epithelium across a filter. Note the spreading of the original sharp peak (in **A**) toward the right; breakdown has resulted in this heterogeneity. The control peak for normal RNA is seen on the right. **C**. RNA extracted from lung epi-thelium after the second 24-hour incubation period. Note that no heavy peak is present in the dense region, in contrast to the situation in **A** and **B**. Instead, the tritium (³H) radioactivity that is present is all in the position of normal, "light" RNA. All this radioactivity had to originate from the mesenchyme; hence transfer of degradation products or of precursor mole-cules probably occurred. (From Grainger and Wessells.)*

supplied to mesenchyme, and if a subsequent transfer of labeled molecules to epithelium is found, how can one establish whether intact macromolecules passed between the tissues? Nucleosides or breakdown products of RNA might yield the same results.

Grainger and Wilt designed a system to circumvent this difficulty. They produced "heavy" nucleosides, ones containing the isotopes ^{13}C and ^{15}N instead of the normal ^{12}C and ^{14}N. RNA that contains the heavy nucleosides is denser than RNA with the normal isotopes. Trace amounts of radioactive (^{3}H) nucleosides and large quantities of the heavy precursors were supplied to chick or mouse lung mesenchyme. After 24 hours, RNA of such mesenchyme was "heavy" and "hot" (radioactive, so it could be traced with monitoring equipment). This labeled mesenchyme was placed on one side of a filter, and unlabeled lung epithelium was placed on the other side. Under the conditions that were used, cellular penetration of the filters would occur (as discussed above) for kidney and cornea. After an additional 24 hours, the mesenchyme and epithelia were removed carefully for separate analysis of RNAs.

The results seen in Figure 14.7 show that no intact "hot and heavy" RNA passed from mesenchyme to epithelium. If it had, then a peak at the high density position should be visible. Instead, a peak of intermediate density is extracted from the epithelial cells. Such molecules contain a mixture of "heavy" and normal nucleotides, a situation that could only result if precursor (or perhaps degradation) products of mesenchymal RNA were transferred to the epithelium. So we can conclude that fewer than 75 molecules of macromolecular RNA per epithelial cell are transferred; and it may be that none at all is transferred.

That is the first conclusion: Under this experimental regime, no macromolecular RNA need move from mesenchyme to epithelium in order for the morphogenetic response to occur. Second, we see that the conventional procedure of simply labeling tissue *A* and measuring isotope in tissue *B* could lead to erroneous conclusions, since abundant transfer might occur, *but not in the form of macromolecules.*

Though Grainger's experiments are exceedingly difficult to perform, the difficulty is the price of a definitive answer. They provide a model for those seeking to demonstrate transfer of "informational" macromolecules between interacting tissues. On the other hand, the fact that there is apparently no exchange of macromolecular RNA leads one to consider other mechanisms for tissue interactions, particularly the cell surface interactions described in this and the preceding chapters.

Kidney development is characterized by reciprocal interaction: the ureteric bud branches in response to nephrogenic mesenchyme; nephrogenic mesenchyme forms tubules in response to the ureteric bud.

Many tissues can stimulate kidney tubule formation.

"Contact" between interacting tissues is required to initiate kidney tubule morphogenesis.

Contact is not required in some other interactions.

Undegraded macromolecular RNA is not transferred from mesenchyme to epithelium during an interaction that leads to lung morphogenesis in culture.

REFERENCES

Kidney development:
L. Saxén. 1971. In *Control Mechanisms of Growth and Differentiation*, Sympos. Soc. Exptl. Biol., 25. Cambridge Univ. Press. This paper summarizes the early literature and provides excellent illustrations of kidney development.

Cell contact across filters:
E. Lehtonen *et al.* 1975. *J. Embryol. Exptl. Morphol.*, *33*, 187. The first clear demonstration of cellular penetration of filters during tissue interactions.

Synthesis in proportion to cell contact:
S. Meier and E.D. Hay. 1975. *J. Cell Biol.*, *66*, 275. This paper demonstrated the linear relationship between amount of cell contact and rates of synthetic response. It includes an excellent bibliography on corneal development that is pertinent to the discussion in following chapters of this book. See also: *Develop. Biol.*, *52* (1976), 141.

Penetration of the basal lamina:
H.C. Slavkin and P. Bringas, Jr. 1976. *Develop. Biol.*, *50*, 428. The latest source of literature on cellular penetration of basal laminae during tissue interactions. In addition, it introduces interactions in teeth, a potentially important system for future studies.

RNA transfer between tissues: 211

R.M. Grainger and N.K. Wessells. 1974. *Proc. Natl. Acad. Sci.*, *71*, 4747. Experiments employing heavy isotope-labeled nucleosides in the lung system.

L. Saxén and S. Toivonen. 1962. *Primary Embryonic Induction*. Prentice-Hall. This scholarly monograph summarized the large amount of work on the "inducer" of the amphibian nervous system.

Chapter Fifteen:

This collagenous layer on the surface of an embryonic mouse lung bronchus is partially stripped from the epithelial surface by mild enzyme treatment.

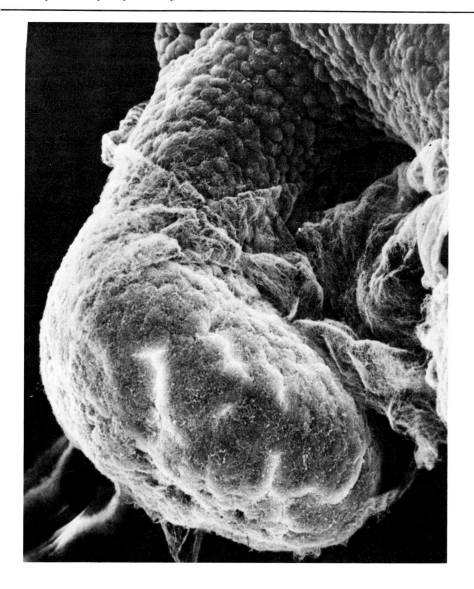

Extracellular Materials and Tissue Interactions

First hyaluronic acid is made in large quantity. Then it is degraded and disappears. What sense does this make in an embryo?

Two classes of macromolecules that are found in extracellular spaces are implicated in various tissue interactions that lead to morphogenesis or cytodifferentiation. These are the collagens and glycosaminoglycans. After discussing some of the characteristics of these substances, we shall examine their roles in interactions.

Collagens

Collagen is the most abundant structural protein in our bodies. Individual molecules are large (about 300,000 daltons), long structures, each composed of three polypeptide chains. Because of the distribution of charge along their lengths, collagen molecules can "self-assemble" into narrow fibrils or wider fibers if conditions of pH, salt concentration, and temperature are appropriate. The result is long, tough fibers

Figure 15.1

The subunit composition of the three main classes of collagen. A single collagen molecule is composed of three polypeptide chains (the alpha units, each with a molecular weight of about 100,000 daltons). Each type of alpha unit (1^I, 1^{II}, 2, 1^{BM}) has a unique amino acid composition; for example, 2 has a greater net positive charge than the other alpha units. The three alpha units of a collagen molecule are entwined in the form of a helix. The collagen molecules can, in turn, associate laterally or end-to-end to form fibers of varying length and width.

Collagen type	Units
Cartilage	$\alpha1^{II}$, $\alpha1^{II}$, $\alpha1^{II}$,
Skin, Bone, Tendon	$\alpha1^{I}$, $\alpha1^{I}$, $\alpha2$
Basal lamina	$\alpha1^{BM}$, $\alpha1^{BM}$, $\alpha1^{BM}$

that provide tensile strength to connective tissues, and serve as major components of cartilage and bone matrices. Alternatively, a finer fibrillar feltwork may form and make up much of the basal lamina near epithelial surfaces. The collagens of basal laminae, dermis, and cartilage differ from each other in actual amino-acid sequences, indicating that separate genes code for the distinctive polypeptides of each collagen type (see Figure 15.1).

Glycosaminoglycans

Glycosaminoglycans (GAGs) are sugar polymers of high molecular weight, composed of repeating dimers of amino sugars (acetylated) and uronic acids (derivatives of sugars such as glucose). GAGs are not generally found as free polysaccharides in the extracellular spaces, but tend to be linked to proteins to form "proteoglycans." Some GAGs are sulfated, and, because they contain abundant free carboxyl groups, are the major anionic substance of the intercellular spaces. Large numbers of GAG molecules of varying length and composition can be linked to the protein of a proteoglycan. By variations in the type and distributions of such GAG, a fantastic diversity of proteoglycans can be generated.

Collagen-GAG

Commonly, collagen is associated with proteoglycans and perhaps other GAG. Important physical properties of collagen change when certain proteoglycans are present. If one considers the possibility of combining the different sorts of collagens with the huge potential number of proteoglycans, it should be clear that a bewildering variety of distinct extracellular macromolecular complexes could be produced.

214

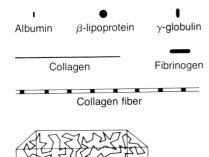

Albumin   β-lipoprotein   γ-globulin

Collagen   Fibrinogen

Collagen fiber

Hyaluronic acid

Figure 15.2
The relative volumes occupied by various proteins and a hydrated molecule of hyaluronic acid (from synovial fluid of a bone joint, tendon, etc.), which has a molecular weight of about 8 to 9 × 10⁶ daltons. The complex folding of the long sugar chain causes the molecule to fill a substantial volume in the extracellular space of an embryo. (Redrawn from T.C. Laurent. In E.A. Balazs, ed., Chemistry and Molecular Biology of the Intercellular Matrix, *Academic Press, 1970.)*

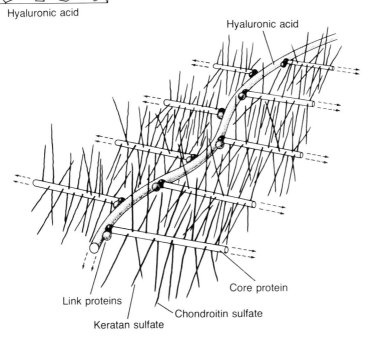

Hyaluronic acid

Core protein

Link proteins

Chondroitin sulfate

Keratan sulfate

Figure 15.3
A tiny part of a proteoglycan aggregate from cartilage matrix. The "core" proteins of each proteoglycan are attached to the hyaluronic acid backbone at intervals of 30 to 60 sugar dimers (uronic acid plus amino sugar) along the backbone. Linking proteins are found at the attachment sites. The GAG molecules (keratan sulfate and chondroitin sulfate) are attached to the core proteins; a given core protein molecule (molecular weight up to 2 × 10⁵) may have 100 sidechains of chondroitin sulfate (molecular weight up to 2 × 10⁴) and 30 to 60 sidechains of keratan sulfate (molecular weight up to 4 or 8 × 10³), and so will be much larger and more complex than the drawing indicates. Since a hyaluronic acid backbone may vary between 4 and 0.4 microns in length, since the length of core proteins is highly variable, and since the length of GAG molecules on the core proteins is also variable and large, the total volume occupied by an aggregate is very great. And, of course, its potential variability in detailed structure is enormous. (From L. Rosenberg et al., and from V.C. Hascall and D. Heinegard, in Slavkin and Greulich.)

215

Collagens in Tissue Interactions

Collagen plays an essential role during development of most tissues and organs. This is demonstrated in several ways, the most satisfactory of which involve interference with collagen synthesis or secretion from cells, or with polymerization of collagen molecules into fibers or proteoglycan complexes.

In developing skin, for example, treatment with the steroid cortisone has a number of effects. One is to interfere with collagen metabolism, with the apparent result that ordered patterns of feathers and scales do not form (see Figure 15.4). Other evidence suggests that an oriented lattice of collagen may appear at the time when dermal cells are forming the early condensations of feather germs and when the feather pattern is being propagated through the skin. The presence of that oriented lattice appears to be a necessary correlate of feather development. This is shown by studies on the chicken mutant "scaleless," homozygotes of which also lack feathers. Interestingly, scaleless skin cells synthesize normal quantities of collagen, but no lattice or

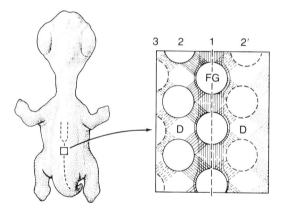

Figure 15.4
The first row of feather germs (1) forms directly on the midline of the lower back. Then rows 2 and 2' appear, with individual feather germs located midway between ones in row 1. Next row 3 (and 3', not shown) develops, with its germs midway between the ones in row 2. In this diagram, the right half represents an early stage, when row 2' is just forming; the left half depicts a later time, when row 3 is forming. The solid, dashed, and dotted lines between feather germs or developing feather germs indicate the direction of tracks of orientation of highly elongated mesenchymal cells in these regions. In contrast, in the areas marked D, mesenchyme cells are not elongated and are oriented at random. The high degree of orientation in the tracks is linked to propagation of the feather pattern over the back, since drugs or mutations that interfere with collagen metabolism or alignment cause a disorientation of the mesenchyme along the tracks that connect prospective feather germs; the result is that mesenchyme remains in the condition characteristic of the D sites, and no feather germs develop.

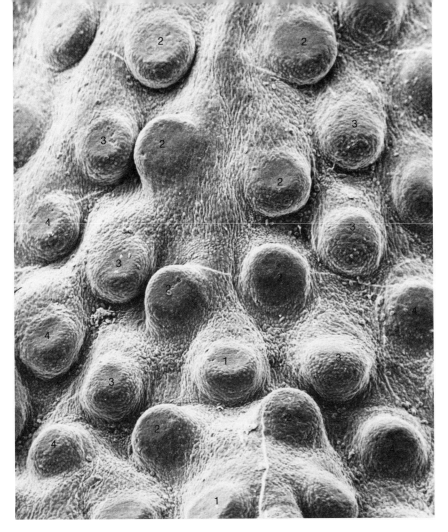

Figure 15.5
A photograph of early feather germs at the "Y"-shaped junction just anterior to the box in Figure 15.4. The row numbers are indicated and the regular spacing of the germs can be appreciated.

dermal condensations appear. Goetinck has shown that the mutation exerts its effect in the epidermis, for in recombinants of normal epidermis plus scaleless dermis, an organized lattice and feathers form; in the converse case, no lattice appears when mutant epidermis is present. It seems likely, therefore, that epidermis exerts some effect on dermis which results in alignment of mesodermal cells, appearance of the lattice, and the subsequent steps in feather and scale development. Note that the requirement is for organization of collagen in a particular pattern, not merely its synthesis and polymerization.

In quite a different sort of developmental process, Hauschka and Konigsberg have found that collagen (derived from a conditioned

218   medium) can stimulate fusion of prospective muscle cells, called myoblasts, into multinucleated skeletal muscle "myotubes," which are the
forerunners of mature muscle cells. Fusion is facilitated by growing the
single myoblasts on a collagen substratum, though other complex
factors may also be required. Since collagen fibers are present in limb
buds and other regions of embryos where skeletal muscle forms, it
seems reasonable to conclude that the extracellular collagen fibers may
play a role in normal muscle development.

At present we may think of the collagens as providing a kind of
scaffolding in developing tissues, either as part of the sheet-like basal
lamina near epithelial cells, or as an intercellular meshwork of fibers.
We have already seen the importance of epithelial cell attachment to
solid substrata for continuance of developmental processes and mitosis. We will discuss below how tracts of collagen and GAG might
serve as routes over which cells move during morphogenesis in embryos. Collagen, as the major structural component of such tracts or
of some basal laminae, may be viewed as a necessary, but not sufficient, factor for the respective developmental processes.

GAGs in Tissue Interactions

A generality has emerged in recent years about one type of GAG,
hyaluronic acid. An early stage of cornea development is characterized
by the presence of an outer epithelium (as discussed above), and underlying primary stroma or mass of collagen and GAG, and an inner
endothelial layer of cells (see Figure 15.6). Toole and Gross found that
at a specific time in development, the latter cells synthesize and secrete
hyaluronic acid into the primary stroma. Presumably, because of the
osmotic properties of these huge polymers, the stroma swells. Immediately, mesenchyme cells located all around the rim of the stroma begin
migrating inward through its swollen interstices. Later, these cells
produce and secrete hyaluronidase, an enzyme that degrades the polymers of hyaluronic acid. Perhaps the hydrolyzed matrix cannot act as
a suitable substratum for cell locomotion, since the mesenchymal cells
soon settle down to the business of producing the secondary stroma.

Several other cases establish the same correlation: cells move
through swollen matrices containing hyaluronic acid; cessation of
movement correlates with appearance of hyaluronidase. If these relationships prove to be truly general for migratory cell populations in
embryos, it will clearly imply that cell movement or nonmovement,
the very basis of stability of tissues, can be controlled by alterations in

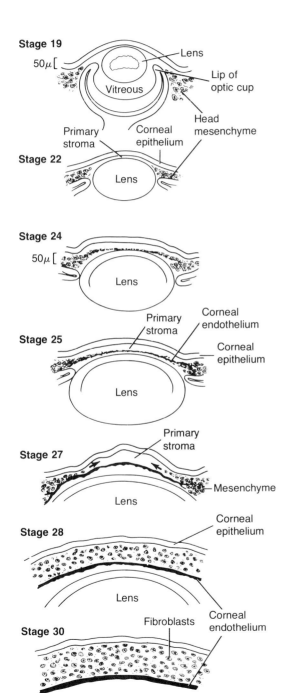

Figure 15.6

Diagrams of cornea development. Between stages 19 and 22 in the chick embryo, the primary stroma begins to accumulate between the lens and corneal epithelium. Then cells migrate along the inner boundary of the primary stroma and give rise to the endothelial layer at stages 25 and 27. Hyaluronic acid content rises in the primary stroma near stage 27, and the peripherally located head mesenchyme cells then migrate into the stroma. Subsequently, hyaluronidase (presumed to come from the endothelium) appears; hyaluronic acid is degraded, and the migratory phase for head mesenchyme ends. (After E.D. Hay and J.P. Revel, Fine Structure of the Developing Avian Cornea. *Karger, 1969.)*

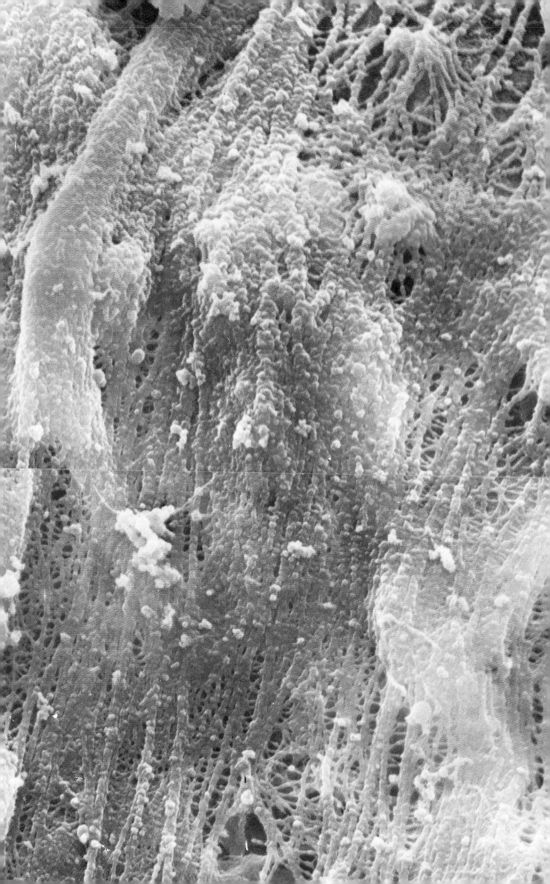

the extracellular environment of cells, not solely by regulation of the locomotory machinery itself.

Other types of GAG exert important effects upon cell differentiation. Notochord, spinal cord, and lens all secrete specific sorts of GAG in the vicinity of various responding cell populations. These responding cells synthesize GAG of the same type that is liberated in their vicinity, and do so at an increased rate (i.e., somite cells do this in response to spinal cord and notochord, corneal cells in response to lens). If GAG is removed from the surface of a notochord by means of enzymes, then the effectiveness of that structure as a stimulant of somite cells is reduced until its extracellular GAG has been resynthesized.

In these cases, GAG acts as a positive feedback signal and increases production of molecules of the same general type. Hay and Meier have speculated whether GAG could also act instructively during the initial interaction between lens and head ectoderm. Thus, the GAG-proteoglycan-collagen complex of the lens surface might serve as the cue to head ectoderm that elicits specific patterns of GAG and collagen production, so that the primary stromal components begin to be made. Thereafter, the lens and cornea become separated by a wide space, but primary stroma constituents could then act as the source of a positive feedback signal which maintains the differentiative activity. Later, the hyaluronic acid/hyaluronidase/cell locomotion phenomena would occur during remodeling of the primary stroma. The hypothesis has an appealing simplicity, and also has the important attribute that it leads to predictions about tissue interactions that are amenable to experimental attack with currently available techniques. Nevertheless, at the moment, there is no evidence showing instructive capability for GAG.

GAG in Salivary Morphogenesis

We now have the background that will allow us to discuss the control of morphogenesis in such organs as salivary gland, lung, or kidney. Recall that localized mitotic activity and intracellular organelles, such

---

◄ **Figure 15.7**
The extracellular material dorsal to a chick embryo's somite. Neural crest cells migrate through and over this meshwork of collagen, upon which GAG has been precipitated with an organic base (cetyl pyridinium chloride). Much of the material seen here may be precipitated hyaluronic acid, the compound often found in regions where embryonic cells carry out movements. (Courtesy of K.T. Tosney.)

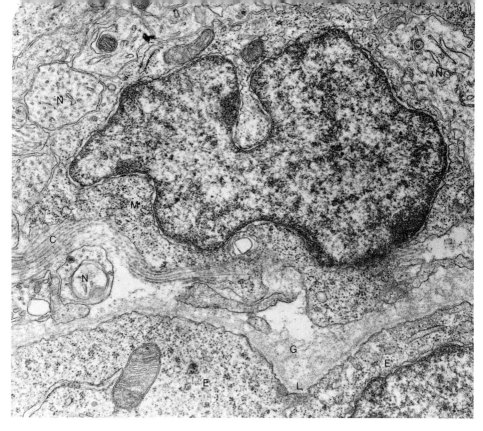

Figure 15.8
The surface of the mouse embryonic salivary gland epithelium, showing typical bundles of collagen (C), the basal lamina (L), and the adjacent extracellular materials that may be largely GAG-containing proteoglycans (G). Many nerve axons (N) occur in this kind of region of epithelial morphogenetic activity. E is an epithelial cell; M is a mesenchyme cell.

as microtubules and microfilaments, may be involved in carrying out morphogenesis in such epithelia. How can we explain the appearance of specific shapes? Why do certain cells form a cleft, while nearby ones do not? GAG, collagen, and remodeling phenomena may provide an answer.

Bernfield and his collaborators first showed the presence of acidic GAG molecules at the interface of salivary epithelium and mesenchyme. If such GAGs, which are about half hyaluronic acid and half proteoglycan, are removed from the surface of an epithelium with enzymes, then the basal lamina disappears and the epithelium "rounds up," losing most or all of its clefts. The epithelial cells, freed from the enclosing basal lamina, bulge outward, form surface protrusions, and undergo redistributions of their microtubules and microfilaments (see Figure 15.9). Only later, after new GAG has accumulated at the epithelial surface and the basal lamina has regenerated, do the cells return to normal morphology. Then, *if mesenchyme is nearby,* clefts form once

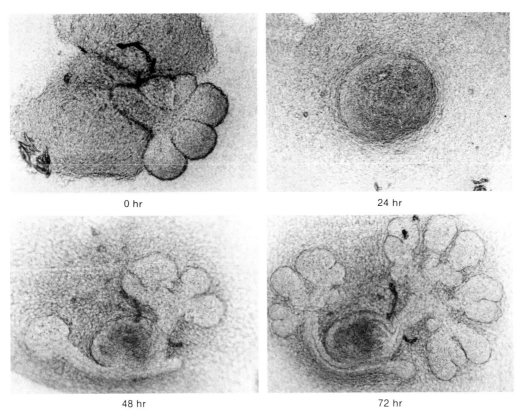

0 hr

24 hr

48 hr

72 hr

Figure 15.9
Effects of treating a mouse embryonic salivary epithelium with enzymes that remove the basal lamina. After the treatment, the epithelium "rounds up" (24 hours), and then reinitiates morphogenesis (48, 72 hours). (Courtesy of M.R. Bernfield.)

more. A major conclusion is that the basal lamina and its GAGs stabilize the shape of the salivary epithelium. We see here still another instance of the importance of a physical substratum to which cells can adhere. This simple phenomenon may be one of the most important in ensuring that cells remain competent to respond to developmental cues and to carry out processes of morphogenesis and differentiation.

Experiments on the molecular basis of salivary morphogenesis provide an interesting lesson on the way that experimental science operates. During the 1960s, the work on the relationship of collagen to fusing myoblasts and to the cornea suggested a crucial role for collagen in organ development. An experiment was performed which suggested that mesenchymally manufactured collagen is deposited near the epithelial surface in the salivary gland system itself. In this experiment mesoderm was exposed to radioactively labeled proline,

224   a precursor of collagen. Then the labeled tissue was cultured across porous filters from unlabeled salivary epithelium. After 24 hours, radioactivity had traversed the filter and was located near the surface of the epithelium, where electron microscopy revealed polymerized collagen fibers. Under such conditions, salivary epithelium underwent a reasonably normal morphogenesis.

The conclusion seemed obvious: collagen might be the essential mesodermal contribution. Grobstein and Cohen (see Bernfield, 1973, for references) then performed the experiment of treating such epithelia with collagenase, an enzyme that digests only collagen proteins. Treated epithelia spread thinly on the culture platform—morphogenesis was inhibited! Next, Cohen and I applied collagenase to lung and to kidney epithelia that had been cultured in the same way, and the results were similar. Electron microscopy revealed that the collagenase removed the basal lamina complex from the epithelium. The complex could reappear if mesenchyme was placed across the filter from the epithelium and the tissues were cultured at 37°C. Presence of the renewed complex was correlated with continuation of morphogenesis. The case seemed conclusive: collagen was crucial for morphogenesis.

Then Merton Bernfield discovered that even the purest preparations of collagenase available, including the types used in the salivary, lung, and kidney experiments, were contaminated with trace amounts of enzymatic activity for digestion of certain polysaccharides. The "collagenase" preparation was, in fact, removing GAG plus collagen fibers, with the resultant rounding-up described above. Subsequent measurements have established the surprising fact that there is little or no collagen in the salivary basal lamina. The accurate conclusion, therefore, is that GAG is the crucial substance required for maintenance of epithelial shape and morphogenesis.

Looking back on the earlier experiments, we can see how a frame of mind had been established in which we were prepared to accept the role of collagen as the critical agent. New techniques, and more critical analyses, allow us to see now the danger of oversimplifying. Of course, we will probably be saying the same thing about much of today's work a decade hence!

Speculations about the Control of Epithelial Morphogenesis

Molecules are contributed to the basal lamina by epithelial cells. An extensive lamina can form on the surface of an epithelium cultured by itself—presence of mesenchymal cells is *not* essential for epithelial cells to carry out such syntheses and secretion. Moreover, the reappearance of normal epithelial cell morphology (see above) as the basal lamina reforms also takes place in the absence of mesenchyme. Neither process is dependent on mesenchyme. Despite that fact, epithelium alone cannot undergo morphogenesis.

The key to the paradox may lie in the dynamic nature of developing systems. Using radioactively labeled precursors of GAG, Bernfield has been able to find out where newly synthesized GAG is deposited or accumulates. Most appears at the tips of the lobes, the sites of rapid cell division and of cleft formation. Much less newly synthesized GAG is found in older, deeper clefts; instead, abundant collagen fibers and a thick layer of "older" GAGs are present there. It looks as if the older clefts are stabilized by these macromolecular complexes.

Prelabeled GAG is lost most rapidly from the morphogenetically active tips. Assuming that this loss reflects "turnover," Bernfield (as

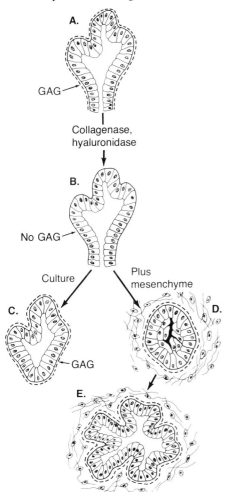

Figure 15.10
*Reappearance of GAG on epithelium. If an embryonic salivary epithelium (**A**) is treated with enzymes, all surface-associated GAG can be removed (**B**). If the epithelium is simply cultured (**C**), those GAG materials appear once again on the epithelial surface since they are synthesized and secreted by those cells. If the GAG-free epithelium is surrounded with appropriate mesenchyme (**D**), GAG reappears as the epithelium rounds up. Morphogenesis and normal branching start soon thereafter (**E**). The synthesis of GAG or collagen is a general property of many developing epithelial populations, and occurs in the absence of living mesenchymal cells.*

226 one possibility) suggests that the mesenchyme may provide hyaluronidase or other hydrolytic enzymes which act to "remodel" such regions and so keep them in an unstabilized condition. Morphogenesis could go on there in the form of cleft formation. Since there is ample evidence for a mesenchymal origin of such enzymes in wound healing, tissue remodeling during metamorphosis, limb development, and cornea differentiation, the idea is not farfetched.

Why do clefts form near regions of GAG turnover? Hyaluronic acid and some other GAGs bind calcium and other metal ions; in so doing the GAG could serve as a source or a trap for such ions. In Chapter 7, we mentioned that certain drugs inhibit salivary epithelial morphogenesis. One of these, papaverine, is known to inhibit contraction of smooth muscle and other contractile events that depend on extracellular calcium ions. Salivary epithelia lose clefts and almost appear to "relax" when treated with papaverine. It is possible, therefore, that extracellular calcium is required by the microfilament-microtubule system of epithelial cells in order to form clefts. GAG, by acting as a kind of ion exchange resin, could influence availability of calcium and so govern indirectly where the morphogenetic events take place. It must be emphasized, however, that the foregoing interpretation is largely speculative.

The reader should now be able to visualize a possible sequence of events that lead to salivary morphogenesis (see Figure 15.12). Epithe-

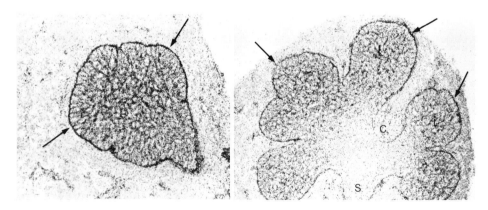

Figure 15.11
Autoradiograms of mouse embryonic salivary glands incubated for 2 hours with ³H-glucosamine, a precursor of GAG. The tiny black dots reflect distribution of incorporated radioactive precursor. Note the accumulation of the dots (i.e., precursor) at the epithelio-mesenchymal interface (arrows); this is also the site of the basal lamina. Much less accumulation of label is seen on the surface of the gland's stalk (S) and deep within clefts (C) than in the morphogenetically active regions (arrows). (Courtesy of M.R. Bernfield.)

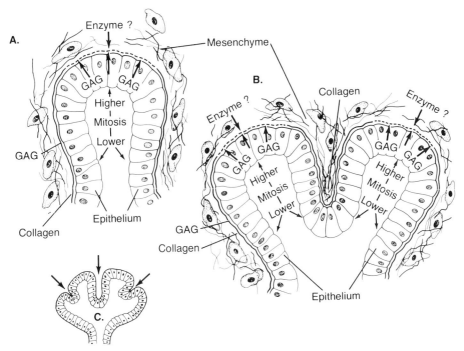

Figure 15.12

*Hypothetical scheme for cleft formation in an epithelium. In **A**, the tip of the epithelium is a major site of new GAG secretion and accumulation in the basal lamina region. Newly synthesized GAG tends to disappear from that tip region, perhaps because of enzyme-induced turnover. Along the stalk region of the same epithelium, the GAG and collagen are more stable, and accumulate in quantity. Mitotic activity is particularly intense near the tip, and that is also the site of cleft formation, perhaps because of microfilament activity. In **B**, the cleft has formed and is stabilized with accumulated GAG and collagen. To each side of the cleft, the tips retain the ability to synthesize, secrete, and apparently turn over GAG. Mitosis and cleft formation occur in those regions (going to the shape outlined in **C**). Though not included in these diagrams, many sites of intimate contact between epithelial and mesenchymal cells occur, and the surface of the epithelium itself is highly dynamic.*

lial cells synthesize and secrete GAGs, mesenchyme cells, collagens. Certain regions are sites of GAG turnover; others are stabilized by GAGs and perhaps by collagen fibers. Intracellular systems, such as microfilaments and microtubules, may act in epithelial cells that are situated near the unstabilized regions, and clefts result. These clefts enlarge and mature by becoming stabilized, but intervening regions continue to be sites of intense mitosis, GAG turnover, and new cleft formation. The tree-like epithelium gradually emerges.

Do these conclusions and hypotheses apply to other organs? Bernfield notes that the main sites of new GAG deposition in lung and kidney epithelia are in the tip portions, where morphogenetic events equivalent to salivary cleft formation go on. Mammary and pancreatic epithelia do not show such distinctive patterns of GAG accumulation,

228   and both undergo quite different patterns of morphogenesis during embryonic stages. Only more observation will decide whether we can apply other conclusions from the salivary system to other branching organs.

Both epithelium and mesenchyme contribute essential molecules to the extracellular spaces in a salivary gland. These molecules may be viewed as essential, but permissive, factors in the tissue interaction. We see no evidence of informational molecules that act instructively. This interpretation is consistent with that of Lawson, who showed that lung mesenchyme can support salivary morphogenesis. Since the GAG distribution pattern of lung and salivary is basically similar, we may guess that lung mesenchyme can contribute the correct types of extracellular factors (remodeling enzymes?) so that salivary cleft formation can go on. Since the shape of the salivary epithelium that develops in lung mesenchyme is normal, it also seems likely that the epithelium governs where sites of turnover and stabilization will be.

If so, how can we interpret those rare cases of mesenchymal influence on epithelial morphology? Recall the mammary mesenchyme that causes salivary epithelium to look like mammary tissue. Could it be that some mesenchymes have limitations on spatial distributions of cells that can produce GAG, collagens, or putative remodeling enzymes, so that an abnormal distribution of stable versus turnover sites occurs on an epithelium? Recall the situation in the lung (Chapter 7) in which bronchial mesoderm stimulates, and tracheal mesoderm inhibits, branching morphogenesis. Perhaps tracheal and bronchial mesoderms provide us with a gross model of more subtle distributions of dissimilar cells in the seemingly homogeneous salivary mesenchyme. Actually, we have no data as yet on this or other possibilities. Nevertheless, we are coming into a position where such hypotheses can be tested and hard answers can be gained.

CONCEPTS

Collagens are large proteins found in extracellular spaces, either as fibers or as part of the basal lamina near epithelia.

Glycosaminoglycans are large sugar polymers, found in the extracellular spaces, frequently associated with certain proteins.

Collagen may be a permissive factor in some types of tissue development, where it may act as a scaffolding or site for attachment of cells.

Given types of GAG may play permissive roles during the expressive phase of development, for instance, during cell locomotion.

In some tissue interactions, a given type of GAG may stimulate responding tissue to produce more GAG of the same type.

GAG is essential for the maintenance of epithelial shape during branching morphogenesis.

REFERENCES

General:
H.C. Slavkin and R.C. Greulick. 1975. *Extracellular Matrix Influences on Gene Expression.* Academic Press. This book presents a wealth of information and references on collagen, GAG, and many of the interacting systems described in this book. Though most of the papers are too short to be of great use, it is a first source for recent references in the field.

Collagen and muscle cells:
D.B. Slater. 1976. *Develop. Biol., 50,* 264. An up-to-date summary of the substratum and nutrient effects on myogenesis, including the Hauschka and Konigsberg papers.
I.R. Konigsberg. 1974. In J. Lash and J.R. Whittaker, eds., *Concepts of Development.* Model experiments and effective criticism of the quantal mitosis hypothesis.

Scaleless and collagen:
P.F. Goetinck and M.J. Sekellick. 1972. *Develop. Biol., 28,* 636. The mutant which is deficient in the collagen lattice and feather development.

GAG and tissue interactions:
S. Meier and E.D. Hay. 1974. *Proc. Natl. Acad. Sci., 71,* 2310. The paper showing effects of GAG on corneal synthesis of extracellular matrix.
E.D. Hay and S. Meier. 1974. *J. Cell Biol., 62,* 889. GAG synthesis in notochord, spinal cord, and lens, all of which evoke similar synthesis in responding tissues.

Hyaluronic acid and hyaluronidase:
G.N. Smith *et al.* 1975. *Develop. Biol., 43,* 221. The latest literature on hyaluronate-hyaluronidase systems, particularly as studied by Toole and Gross. This paper deals specifically with those agents in regeneration of limbs.

Salivary gland morphogenesis:
M.R. Bernfield *et al.* 1973. *Amer. Zool., 13,* 1067. A review of GAG in salivary development. Recent work will appear in *J. Cell Biol.*

Chapter Sixteen:

The surface ectoderm of a salamander embryo, showing the remarkable ciliated cells that are scattered at near-equal distances over the surface of the embryo. It is not known how this regular spacing is maintained as the embryo and its surface grow during development. Similarly, it is not clear how ciliary beat is coordinated among these seemingly separate cells, so that the cilia of all cells in a given region of the embryo beat in a characteristic direction. The cell to the right of center is also seen in the frontispiece to Chapter 1. (Courtesy of P. Karfunkel.)

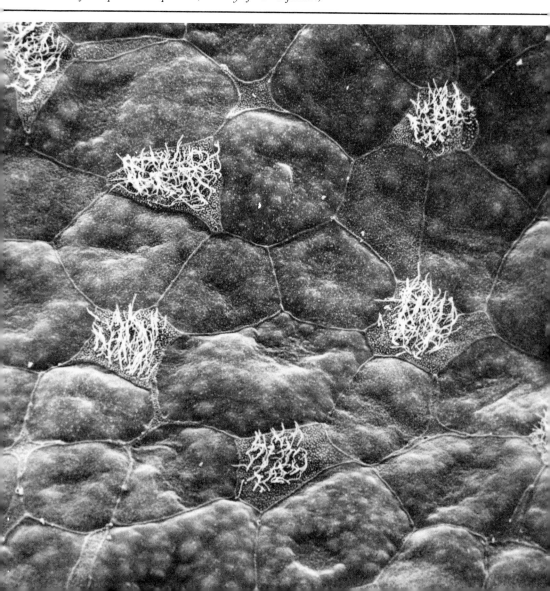

Tissue
Versus Cell

The number of cells may not matter very much for many parts of development.

An issue we have touched on repeatedly, but never addressed explicitly, is the relationship between the individual developing cell and the cell population—the tissue—in which it resides. We believe that the restricted state is a property of single cells—it is *heritable;* so its stability must be regarded as a cellular phenomenon. In contrast, expression is a function of the cell population and the surrounding environment—its stability is a population phenomenon.

Suppose we ask the question: What is the origin of restriction? On reconsidering the experiments, we realize that what has been tested is interactions between *tissues.* Not once has a case been cited in which cell *A* acts on cell *B*, or tissue *A* on cell *B;* current techniques limit us to examining how *tissue A* affects *tissue B.*

But the restricted state is a cellular property. A major question, therefore, is whether instructive tissue interactions act on a responding population as a *tissue* (setting up conditions which lead to restriction

232   of individual cells), or whether they act directly on individual cells (e.g., does the optic vesicle cause restriction of individual lens cells?).

We can appreciate the significance of our dilemma by the following example. Suppose we cut away and discard one half of an early amphibian embryo's limb bud or half of the medullary plate tissue that would form the optic vesicle. The wounds heal. Development proceeds. A complete limb or a complete eye forms. Cells that normally would have given rise to only half a limb or half an eye have yielded the complete organs. A process of "regulation" has occurred. The cells are apparently affected by what the British biologist Lewis Wolpert (1971) has called *"positional information."* This is, of course, a new term for a long-recognized set of phenomena, ones that have been studied since Harrison adapted Born's transplantation procedures and began the experimental attack on morphogenesis. In the limb bud that we cut in half, the remaining cells would (for example) normally have given rise to the posterior half of a limb. However, when the anterior half-bud is missing, the *relative position* of remaining cells is altered, and some behave in a way that is appropriate for "anterior" cells.

Another intriguing illustration of positional information is described in Figure 16.1. In this and several other experimental arrangements, it is clear that regenerating salamander limb cells behave as if they can "sense" which way is proximal, and which is distal; the result is that they always regenerate the "distal" parts.

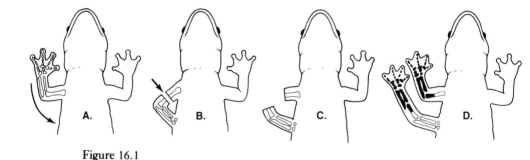

Figure 16.1
Position and limb regeneration. First, the hand portion of a salamander's forelimb is inserted through the lateral body wall and allowed to heal in place. Then the original upper arm (humerus) region is cut **(B)**. *Regeneration occurs from both stumps if both have adequate nerve supplies. The front one forms the expected proximodistal sequence of missing parts. The rear stump does not regenerate its original missing part—the proximal half of the upper arm (humerus). Instead it regenerates a normal proximodistal sequence of arm parts: half humerus, radius-ulna, wrist, digits. The regenerating system "behaves" as if it can sense which way is toward the body trunk and which way is distal. The information used in that sensing and decision making is termed "positional information" by Wolpert. (Experiments by Butler; redrawn from R.J. Goss,* Principles of Regeneration, *Academic Press, 1969.)*

These kinds of positional effects can even be demonstrated for
developing insect imaginal discs, structures that are notorious for
precocious restriction of their component cells. These positional effects are the essence of how tissue-level information controls cell-level
behavior. Positional information is clearly a primary means of integrating cells into tissues and tissues into the whole embryonic organism. It is crucial, therefore, to define positional information in terms
of physicochemical realities, so that we do not have to depend on a
nebulous concept to "explain" these important processes in embryos.

Coupling of Cells

The dilemma of distinguishing cell from tissue properties in development is compounded by the phenomenon of "low-resistance coupling"
between cells. Suppose we insert the tip of a microelectrode into
epithelial cell A (Figure 16.2), and place another electrode in cell B,
located well away from A in the cellular sheet. If we then pass current between the electrodes, we find that the electrical resistance of
the paths differs dramatically: A to B has a low value; A to ground
has a high value. In other words, ions can flow easily through the
epithelium from one cell to the next, but they cannot flow easily from
inside the cells to the extracellular space; "insulation" intervenes on
that pathway.

It is concluded that cells in an epithelium can be connected by low-
resistance junctions. Electron microscopy reveals specialized "gap"
junctions between cells that exhibit low-resistance coupling. Perhaps these junctions are the sites of current passage (see Figures 16.3
and 16.4).

What is the significance of the coupling phenomenon? If we inject molecules of low molecular weight (to 500 daltons) into cell A,
we soon observe that the substance diffuses rapidly from A through
the epithelial population. The cells are in fact freely permeable to

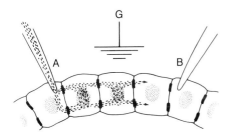

Figure 16.2
*A population of embryonic cells connected
by gap junctions. Electrical current (or
small molecules) can pass freely from
electrode A to electrode B, but not from
A across the cell surface to ground, G.*

234   many substances up to 500 daltons in molecular weight! The ion flow used to demonstrate low-resistance electrical coupling merely reflects this high permeability (note that some electrically coupled systems are *not* permeable to small molecules). What is important here is that cells are not behaving as isolated units; they are truly integrated into a higher-order functional unit, the tissue.

Think of the consequences of this conclusion. Changes in metabolism or in the sizes of pools of small molecules in one cell can affect nearby cells; in other words, cells can be metabolically "coupled" (Gilula *et al.*, 1972). Factors acting on one cell can exert indirect effects on neighboring cells. Only at the level of macromolecules—proteins, nucleic acids, etc.—are coupled cells truly isolated from one another.

There is no clear or constant relationship between low-resistance coupling and cell behavior. Thus normal cells may or may not establish gap junctions; likewise, transformed cells may or may not possess the junctions and coupling. The situation for tissue behavior is equally confusing, in that some epithelia display coupling, but others do not. Therefore, the phenomenon cannot be linked in a straightforward way with either instructive or permissive tissue interactions. However, Dixon and Cronly-Dillon have described an important case involving an expressive event. At a certain time, nerve cells of the eye's neural retina stop dividing and become permanently "fixed" on the retinal map (i.e., they assume an "address" or predisposition to connect at a specific locus relative to neighbors on the optic tectum of the brain). At the same time, the cells lose their gap junctions! Instead of being integrated into the epithelial population, they are now isolated individuals. How this relates to the supremely important (for a nerve cell)

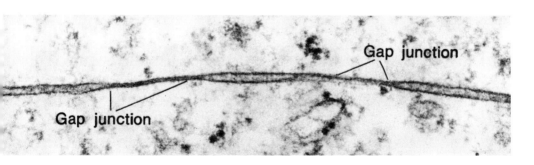

Figure 16.3
"Gap" junctions between two cells in the testis of a rat. Note the close apposition of the plasma membranes of these cells in the junction region. It is here, in the regions of closest apposition, that the aggregations of intramembranous particles (such as in Figure 16.4) are found. These gap junctions may also be the site of low-resistance "coupling" and metabolic "coupling" between cells (see text). (Courtesy of D.W. Fawcett. From Develop. Biol., 50 *(1976), 142.)*

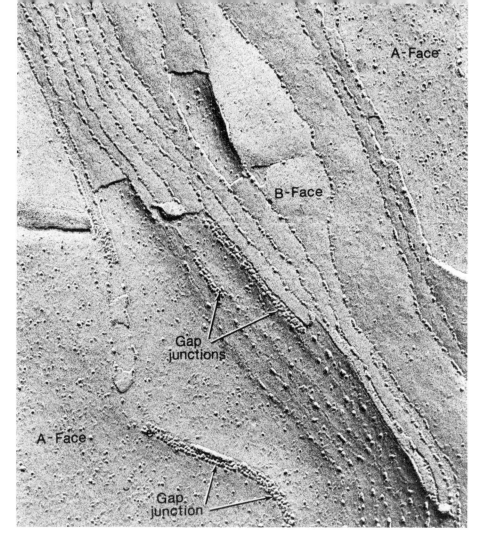

Figure 16.4
Intramembranous particles in the surface membranes of Sertoli cells in the rat testis. Gap junctions are characterized by densely packed aggregates of such particles, whereas normal plasma membrane of the cell surface (signified here by "A-face," "B-face") have more widely scattered particles. The particles are made visible here by means of a technique that involves freezing of the cells and fracturing of their surfaces through the plane of the plasma membrane. (Courtesy of D.W. Fawcett. From Develop. Biol., *50 (1976), 142.)*

property of becoming positionally specified in the eye is an intriguing mystery.

Despite the contradictions and difficulty of interpretation in this area, the speed with which gap junctions can form (of the order of seconds or minutes, not hours or days), and the potential consequences if they do form, both argue for the importance of further investigations. The results will impinge directly on the problem of distinguishing cell-level from tissue-level activity in development.

Abnormal Cell Numbers and Development

Both in this and in earlier chapters, the significance of cells or groups of cells for continuance of normal development has been implicitly assumed. Two sorts of experiments complicate the view that specific numbers of cells are required for given developmental events. First are observations by Dennis Smith (Smith and Ecker, 1970) in which eggs of the frog *Rana pipiens* were removed from a female's ovary and were treated briefly with progesterone, a female sex hormone, to initiate the terminal egg maturation processes. The eggs were then allowed

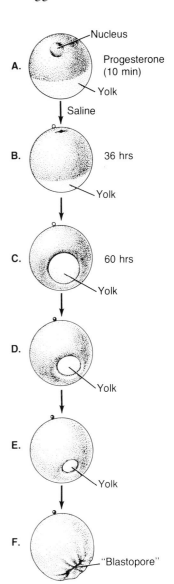

Figure 16.5
Pseudogastrulation in frog eggs. An egg is dissected from the ovary, exposed to progesterone for 10 minutes (A), washed, and incubated in a normal dilute saline solution. Meiosis (the reduction divisions) proceeds to midway through the second division by 36 hours (B). About a day later, the more darkly pigmented portion of the uncleaved egg's surface begins to cover the whitish, yolky area (C). The yolky area sinks inward just as it would during gastrulation in a many-celled embryo (D, E). Ultimately, the edges of the dark cell surface meet, leaving a slit-like opening, analogous to the "blastopore" (F) of a normal gastrula-stage embryo. If the nucleus is removed from the dissected egg prior to hormone treatment, then the pseudogastrulation movements initiate anyway and proceed to the stage seen in D; apparently the information required to carry out this morphogenetic movement resides in the cytoplasm of the mature egg. (After Smith and Ecker.)

Figure 16.6

Unfertilized but hormone-activated frog eggs seen near the time when pseudogastrulation is complete. The white yolky cytoplasm has completely disappeared from the surface of each single cell as the "blastopore" appears to close. (Courtesy of L.D. Smith. From Develop. Biol., *22 (1970), 626.)*

to sit undisturbed in a physiological salt solution (no sperm were added or other means used to activate the eggs). Approximately 2.5 days later, a portion of the egg surface began to sink inward as if gastrulation were occurring (see Figures 16.5 and 16.6). And in fact a blastopore-like groove and "yolk plug" stage were seen. This "pseudogastrulation" phenomenon is accompanied by internal cytoplasmic movements that resemble those carried out by the population of invaginating cells in the normal gastrula. It is interesting that at least the early phases of pseudogastrulation occur even if the nucleus—the so-called germinal vesicle—is removed prior to progesterone treatment. Furthermore, other experiments demonstrate that progesterone apparently triggers these developmental maturational events by acting *at the cell surface.* That is, injected progesterone does not activate eggs; only bathing the eggs in a solution of the hormone allows the activation to start. These results suggest that much of the business of the crucially important gastrulation process is programmed in the cytoplasm and at the cell surface (recall too the "o" factors in salamanders, Chapter 2). It is almost as if compartmentalization into individual cells is incidental to the use of this program of information and to the actual mechanics of gastrulation. This is an exaggeration, of course, since pseudogastrulation is not normal or complete. Nevertheless one can see that a complicated developmental process, normally attributed to a unique

238 population of early gastrula cells (those derived from the gray crescent region), is a consequence of the activity of cytoplasmic substances that happen to be segregated into those cells as a result of cleavage divisions.

A related type of time-dependent morphogenesis is seen in early mammalian embryos in which cleavage is arrested when only a few cells (e.g., 2, 4) are present. Despite this small number of blastomeres, a hollowing out or vesiculation occurs to yield a blastocoele-like cavity (the blastocoele is the cavity within a blastula-stage embryo; in mammals, it is the cavity within the trophoblast). This cavity formation occurs at about the same time that it would in unarrested normal embryos composed of considerably more cells. So, like pseudogastrulation in frog eggs, a morphogenetic process is carried out on schedule without seeming regard to cell number.

Other kinds of observations raise questions about the significance of cell numbers in development. Among various amphibians, widely differing degrees of ploidy are found in natural populations or can be generated in the laboratory. Salamanders of the genus *Triturus* may have haploid, diploid, triploid, and on up even to hexaploid, chromosome sets in each of their somatic cells. Developmental abnormalities are seen in higher ploidy. Nevertheless, in the less extreme degrees of ploidy, many larvae and adults develop with a full complement of differentiated cell types (and organs).

What can be learned from these creatures? First, we note that the size of the whole pentaploid embryo or larva is normal. Next, suppose we concentrate on an internal organ, such as the kidney, and examine its structure. Again, the size of the organ is normal, as is the circum-

Figure 16.7
Polyploidy and number of kidney tubule cells. Here are cross sections through kidney tubules of haploid, diploid, and pentaploid amphibian larvae. Note that the number of cells goes down as size and number of chromosome sets go up. In the pentaploids, a single cell may wrap much of the way around the whole tubule; the shape of that huge, curved cell is quite different from that of the smaller, flatter haploid cell in the wall of a tubule. (After Fankhauser. From J. Exptl. Zool., 100 *(1945), 445.)*

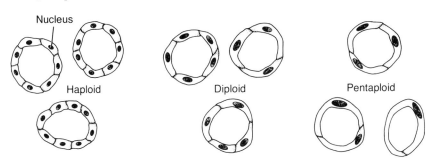

ference of the individual tubules in nephrons. However, when we inspect a cross section of such a tubule, a novel situation is revealed. Only one or two huge, curved cells form the wall of the tubule. In haploids and diploids of the same species, nine and five cells, respectively, might make up the wall of a similar tubule. As seen in Figure 16.7, the smaller numbers of cells in high-ploidy tissues must each assume a shape somewhat different from that of similar cells in a diploid. If this did not occur, the normal dimensions of the tubule or the kidney (as an organ) could not be attained.

Several generalizations have been pointed out by the Princeton biologist, Gerhart Fankhauser.

1. There is control of organism and of organ size in development.
2. There is no apparent control of cell number and only marginal control of cell shape for a given organ.
3. Cell size is directly proportional to ploidy; the larger the number of chromosome sets, the larger the cell.
4. Cell size and cell number are inversely correlated; as size increases, numbers fall.

With these "rules" in mind, we can understand the pentaploid kidney. Very large cells are present because of the high ploidy. The organ and its nephrons are built of a small number of building blocks, the cells, but they achieve a final, functional condition (because of sufficient malleability in cell shape) in which the small number of large cells can yield normal tissue and organ shapes.

Essentially the same conclusions can be reached from a very different experimental procedure. Cooke has applied the drug colcemid (a derivative of colchicine) to prevent cell division in embryos of *Xenopus laevis*, the African clawed toad. If this is done when there are about 5,000 cells present, gastrulation proceeds, an early nervous system develops, and other tissue types differentiate. Untreated embryos are composed of about 28,000 cells when such processes are completed. Here, as with increased ploidy, small numbers of large cells can carry out the expressive events of differentiation and morphogenesis.

Next, Cooke implanted a piece of invaginating, mesodermal tissue into a host embryo in order to test whether tissue interactions could occur in the absence of cell division. Despite the presence of colcemid and attendant inhibition of mitosis, an extra set of tissues developed from the host cells. This implies that both restrictive and expressive events can go on in host cells where there is neither normal cell division nor the accumulation of normal numbers of cells. Unfortunately, this case is somewhat complicated by the report that DNA synthesis is

240   not halted in the cell division-arrested embryos. Whether the colce-
       mid cells are becoming polyploid, or what, is not known. Never-
       theless, this observation means that the result on colcemid-treated
       embryos does not necessarily contradict the conclusions on pancreas,
       oviduct, and other cell systems, discussed earlier, that imply DNA
       synthesis is required for certain types of regulatory transitions to
       occur.

           These experiments, the ones on increased ploidy, and those on
       pseudogastrulation, all provide strong warning that it is premature
       to assume that specific numbers of cells, of cell-division cycles, or of
       other variables of population dynamics are *necessary*, and not *incidental*,
       parts of tissue and cell interactions in development.

Cell Number, the Cell Surface, and Gene Activity

To conclude this chapter, we will consider observations that link the
previous discussion of cell surface (Chapters 2, 13) with our current
discussion on cell number and gene functions.

Figure 16.8
*A young larval tadpole of an ascidian. Note the small number of cells, and the regular spacing
and shape of cells in the tail of this minute creature. (Courtesy of J.P. Wourms.)*

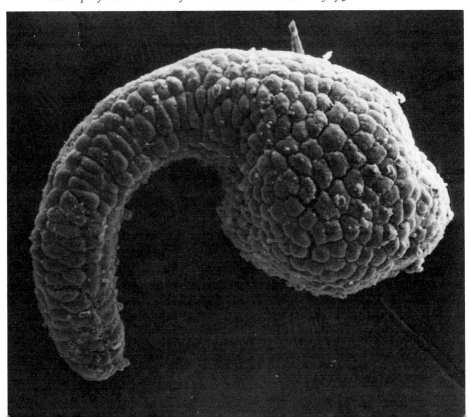

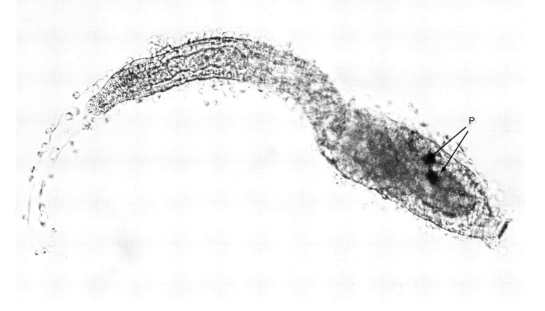

Figure 16.9

A late tadpole of the ascidian Ciona intestinalis. *Note the two pigmented cells (P) in the head of the tadpole. They are the cells that arise as a result of the peculiar lineage behavior described in the text. The tail of the tadpole seen here was in the process of being resorbed during metamorphosis of the free-swimming tadpole into the sessile, adult-type organism. Microfilaments similar to those used during morphogenesis of epithelial organs are involved in the rapid shortening of the tail.*

Ascidians are primitive chordates, marine organisms whose embryos develop rapidly and carry out a precisely ordered cleavage process in which areas of the zygote cortex are allocated to specific blastomeres. Since important developmental events occur in them when only a few blastomeres are present, we can trace the exact lineage of many differentiated cell types. One lineage forms the muscle cells in the tail of the tadpole larva. These differentiated muscle cells contain the enzyme acetyl cholinesterase, which is the enzyme used to hydrolyze the neurotransmitter acetyl choline at motor end plates (Chapter 11).

Another cell lineage gives rise to just *two* individual pigment cells located in different parts of the larval brain (see Figure 16.9). The way that these cells arise is interesting in light of our earlier discussion of stem cells and taste buds (Chapter 11). At the 32-cell stage in embryos of the genus *Ciona*, two bilaterally placed blastomeres can be identified as the ancestors of the two pigment cells. Each of these ancestral cells divides three times. The surprising thing is that one daughter cell that arises from each of these divisions *lacks* the ability to form pigment, whereas the other daughter retains it. After the third division, the single cell in each lineage that retains this ability ceases mitosis, forms the enzyme tyrosinase, and uses that enzyme to manufacture the dark pigment melanin.

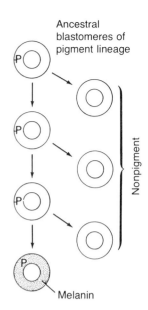

Figure 16.10
A pigment lineage. At the 32-cell stage in a Ciona *embryo, two cells give rise to the separate pigment lineages. The mitotic behavior of one lineage is seen here. Only one daughter cell of each division retains the ability to form pigment. One way that the precision of this process could be maintained is indicated: a "propigment factor" (P), perhaps derived from the egg cortex, is fixed to a portion of the cell surface. The cell line that retains that cytoplasmic information ultimately can differentiate into a pigment cell. (After Whittaker.)*

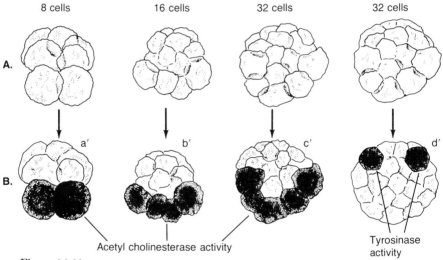

Figure 16.11
Enzyme activity in cleavage-arrested ascidian embryos. At the stages shown on line A (8, 16, 32, and 32 cells), these embryos were placed in a drug that inhibits cleavage divisions (cytochalasin B, a substance produced by certain fungi that interferes with many processes that depend on microfilaments and cell surfaces). Cleavage stopped at the stage indicated. Then, several hours later, and at the same time they would occur in untreated developing embryos with many more cells, the two enzyme activities appear. Chemical reactions are carried out on the intact embryos in such a way that a visible reaction product accumulates at the site of enzyme localization. It is seen here that acetyl cholinesterase is restricted to 2 of the 8 cells in embryo a'; to 4 of 16 in b'; and to 6 of 32 in c'. Other kinds of studies show that these cells are the lineage that gives rise to muscle cells. Embryo d', which was arrested at the 32-cell stage, shows the tyrosinase reaction in just two bilaterally placed blastomeres. These cells are the lineage cells described in Figure 16.10 that give rise to the two pigmented cells of the larva. (Tracings from J.R. Whittaker, Proc. Natl. Acad. Sci., 70 (1973), 2096.)

Whittaker has interfered with the cleavage divisions of *Ciona* embryos by using a number of drugs. He finds, however, that both
cholinesterase and tyrosine activities appear *on schedule*, and *only in
the ancestral cells of the muscle and pigment lineages* (see Figure 16.11).
Thus, in an embryo arrested at the 32-cell stage, the six prospective
muscle cells develop cholinesterase and the two prospective pigment
cells (of the separate lineages; see above) develop tyrosinase and melanin
at the usual time. The normal number of division cycles does *not* have
to occur, nor do a specific number of blastomeres need to accumulate
in order for the genes to be expressed.

Both actinomycin D, the inhibitor of RNA synthesis, and puro-
mycin, an inhibitor of protein synthesis, prevent activity of the two
enzymes from appearing. This tells us that the transcription and translation processes are being activated on schedule in the cleavage-arrested
cell lineages. What is the source of the information that activates the
genes? Whittaker points out that cell lineages appear and development
occurs normally in many embryos that have been centrifuged to rearrange cytoplasmic components, or even in ones from which much
cytoplasm has been removed. The portion of the cell that is not altered
by centrifuging or removal of cytoplasm is the cell surface—the plasma
membrane and underlying cortex. And, of course, it is that same cortex
which is segregated so precisely during cleavage.

Here we have an important general model for tissue interactions.
In *Ciona* eggs, the distribution of cell cortex components may be governed by interaction of the developing egg cell with surrounding ovary
cells in the maternal organism (see Figure 16.12). Proximity of an
ovarian follicle cell to a portion of the egg surface might allow: (1) transfer and localized deposition of informational substances; (2) localized
deposition of informational substances that are manufactured in the
egg itself; or (3) modification of that portion of the cortex so that, following the cytoplasmic rearrangements which result from fertilization
(Chapter 1), informational substances bind to that portion. The important concept is that interaction between cells could affect the distribution of developmental information within a responding cell.

The subsequent parceling out of that information accomplishes
two things. First, it controls the relative positions of cell lineages, and
so, indirectly, the kinds of interactions that these lineages will be exposed to later. Second, it actually provides the cytoplasmic information that is necessary for gene activity as the progeny of the lineages
differentiate.

Since the two enzyme activities appear at the correct time after
fertilization (recall pseudogastrulation occurring on schedule in am-

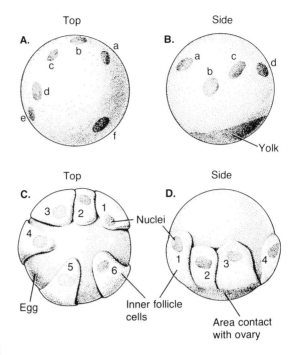

Figure 16.12
Relationship between maternal cells in the ovary and cytoplasmic components of the egg. Top
(A) and side (B) views of a snail (Limnaea) egg are shown, including six regions of cortical
cytoplasm that have unique staining properties with certain dyes. The main mass of yolk is
located toward the lower pole of the egg in B. In C and D, top and side views of the same egg
are seen, but with the placements of the ovarian follicle cells indicated. Note that the nuclei
of the follicle cells (1-6) are situated directly over the sites in the egg cytoplasm that will have
the distinctive staining properties in a mature egg (a-f). In addition, the main area of contact
with other ovary cells is shown; it corresponds to the yolk region of the mature egg. This
would not be surprising if much of the yolk is transferred to the egg cell from that part of the
ovary. Although the developmental significance of the staining patches in the egg cortex is
unknown, their obvious spatial correspondence to the follicle cells provides us with a model of
how ovary-egg interaction could affect distribution of cytoplasmic information in eggs. (Trac-
ings from C.P. Raven, Develop. Biol., *7 (1963), 130.)*

phibian eggs), some type of "counting mechanism" must operate,
though it cannot depend on numbers of cell or of nuclear divisions.
This conclusion is similar to that drawn from the ploidy studies and
Cooke's mitosis-arrested amphibian embryos. It gives one pause in
considering the significance of the timing of cell division cycles during
the progress zone function in limbs (Chapter 6; recall that positional
values are assigned quantally in periods that correspond to division
cycles). Whittaker speculates that the "clock" effect in *Ciona* blasto-

meres may be due to a time-dependent release (from the cell cortex) of the informational molecules that lead to cell differentiation and activation of gene activity.

In more general terms, these experiments suggest several possibilities. Could some instructive and permissive tissue interactions involve action at the surface of responding cells, either by the addition of informational molecules or by the modification or release of ones already there? Suppose, for instance, that each cell in the lineage derived from the gray crescent region of amphibian zygotes has cortically located substances that could cause muscle, cartilage, bone, dermis (and so forth) patterns of differentiation in those cells. An interaction leading to the release of one type of information might yield muscle development, and so on. Obviously, any number of models like this could be imagined. What is important is that we should look to the cell cortex as a site and source of information for cell development.

CONCEPTS

Many developing cells behave as if they can "sense" their position in the cell population or region of an embryo.

Some cells may establish specialized junctions with other cells that are able to serve as points of exchange of ions and small molecules. In such cases, and for certain phenomena, the population is the "unit," not the cell.

A process remarkably like the early gastrulation movements can occur in some hormone-treated amphibian eggs which do not cleave into separate blastomeres.

By causing increases in ploidy, or by inhibiting mitosis, abnormally small numbers of cells can be shown to be able to carry out much of normal development.

The actual number of cells does not appear to be a crucial factor in many developmental processes.

Materials that are segregated to specific cell lineages during cleavage can exert effects in a time-related manner on cellular differentiation and gene expression.

REFERENCES

Growth control in general:

R.J. Goss, ed. 1972. *Regulation of Organ and Tissue Growth*. Academic Press. A series of papers dealing with many important aspects of growth control in a variety of organisms, including mammals.

Positional information:

L. Wolpert. 1971. *Current Topics Develop. Biol.*, *6*, 183. This is the classic paper in the field and presents the formal analysis of the problem.

C.E. Wilde. 1974. In J. Lash and J.R. Whittaker, eds., *Concepts of Development*. Sinauer. A more down-to-earth, but thought-provoking essay emphasizing the dimension of time in relation to positional information.

V. French, P.J. Bryant, and S.V. Bryant. 1976. *Science 193*, 969. This paper proposes the best explanation so far offered for "positional" information.

Limb reversal and distal outgrowth:

E.G. Butler. 1955. *J. Morphol.*, *96*, 265. The classic experiment in which a limb tip is inserted in the side, distal amputation is performed, and the unexpected result occurs!

Low-resistance coupling:

R. Cox, ed. 1974. *Cell Communication*. Wiley. Two papers, by J. Sheridan and by N.B. Gilula, summarize current views of cell surface junctions, coupling, and metabolic exchanges between cells.

J.S. Dixon and J.R. Cronly-Dillon. 1972. *J. Embryol. Exptl. Morphol.*, *28*, 659. This paper demonstrates the loss of coupling when neural retina cells become "specified."

N.B. Gilula *et al.* 1972. *Nature, 235*, 262. The major paper pointing out the consequences of coupling for metabolic pathways.

Pseudogastrulation:

L.D. Smith and R.E. Ecker. 1970. *Develop. Biol.*, *22*, 622. Observations on hormone-treated but unfertilized eggs.

Ploidy variations and cell numbers:

G. Fankhauser. 1955. In B.H. Willier, P. Weiss, and V. Hamburger, eds., *Analysis of Development*. Saunders. This paper gives references to many of the original papers in the field; see in particular, *Quart. Rev. Biol.*, *20* (1945), 20.

Mitotic inhibition and development:

J. Cooke. 1973. *J. Embryol. Exptl. Morphol.*, *30*, 49; also *Nature, 242* (1973), 55. These papers show that much of early development can proceed normally in the absence of mitosis.

Segregation and release of cytoplasmic information:
J.R. Whittaker. 1973. *Proc. Natl. Acad. Sci.*, 70, 2096. This paper includes the work on cholinesterase and tyrosinase, and presents an excellent summation and bibliography on the phenomena.

Chapter Seventeen:

Nerve axons growing in culture from the parasympathetic ganglion of a mouse embryo's salivary gland. This intricate interweaving gives but the barest hint of the fantastic complexity of axonal networks in intact nervous tissue. (Courtesy of M.D. Coughlin.)

Multicellularity, Integration, and Tissue Interactions

Types of information vary during the life cycle.

The significance of tissue interactions in the life cycle of an individual organism can be seen if we extend our discussion to the context of evolution.

One consequence of the origin of multicellularity was the appearance of a "division of labor" among the cells of an individual. Ultimately, the condition we see today—many different cell types performing discrete functions in the single organism—evolved. Pancreas, lens, or even free-floating white blood cells cannot survive as independent organisms. All derive support from the other cell types that compose the organism.

The key to survival of the diverse cell types is *integration*, the process wherein discrete cell populations function in a coordinated manner so that the whole organism can survive to reproduce. Two major sources of this coordination in adults are the nervous and the endocrine

250   systems. Obviously this is true for control of gross behavior and regulation of basic physiological processes. It may also be true of specific trophic factors that help maintain the functional, expressed state of cells. Similarly, so-called "inhibitory" regulatory loops, like the lens iris system or chalones, may be equally essential in controlling cell population size and stability of the expressed state.

The embryo, too, must be integrated. First, it must be integrated on an ongoing functional basis (witness development of special embryonic membranes such as the amnion or the yolk sac, or appearance of temporary ciliary swimming organelles). Second, its development must be integrated. In other words, developing cell populations must integrate in such a way that discrete cell types arise, organs are located in correct positions, and differentiation (expression) can go on. Positional information and tissue interactions are the core of this embryonic integration.

For largely unexplained reasons, only a limited amount of developmental information is built into the mature egg. Certain developmental processes are set in motion as the new nuclei become associated with discrete cytoplasmic regions—the apparent sites of early developmental information. Then, in a process that is universal among multicellular animals, cell populations are rearranged in space to permit the interactions between dissimilar cell types that initiate the next set of developmental events. Ultimately, the stage is reached in which the integrating control of nerves, hormones, inhibitors, and growth regulators operates.

Thus we might envision three levels of information essential to integration: first, that built into the egg; second, an intermediate type, characterized by tissue interactions in embryos; and third, the nervous, endocrine, and other regulatory systems in postembryonic life.

But why not build all required developmental information into the DNA itself? The neurophysiologist Donald Kennedy suggests an apt analogy: the DNA is like a "prime contract"—the plan for the building blocks and some basic rules or limitations for the way that the job can be performed (the regulatory genes). A series of "subcontracts" actually determines what is built. Materials in the egg cytoplasm and their distribution, interactions between tissues, and analogous events use the prime contract, but do so with a considerable degree of freedom. If the carpenter holding a subcontract places a wall somewhat off center, the electrical subcontractor does not string wires in the air from ceiling to floor; the subcontractor adjusts for the earlier errors and puts the wires in the wall. So, too, developing nerve axons get to the "right" spot despite abnormalities that lead a limb to be located incorrectly. This sort of adjustment is *not* possible at the prime contract (DNA) level (except, of course, as the result of mutational processes, which are irrelevant here). But adjustment, response to the changing environment within an embryo, is the *basic feature* of cell and tissue development—witness positional information and the halved limb bud!

It may be useful, then, to think more about Kennedy's analogy. The DNA, as a code for determining primary protein structure, is a peculiar kind of prime contract, since it really tells how to make bricks, mortar, and pipes, not a building. There is no picture, no blueprint of the final product included in the DNA. It is only by use of other types of information that the building blocks are put together correctly in space and time, and the building itself gradually emerges. Thus, positional information, tissue interactions, and other integrating factors, by controlling and coordinating the use of the basic information in DNA, generate development of the multicellular animal organism.

CONCEPTS

Tissue interactions are one of the sources of information available to developing cell populations.

Tissue interactions are one of the means by which coordination between developing cell populations is maintained, and by which at least a limited capacity for correcting errors is provided to the embryo.

Chapter Eighteen:

Desert butte? No, a late stage in the development of a scleral papilla near the cornea of a chick embryo *(see Chapter 4)*. A scleral bone formed directly beneath this mound of epidermal cells in response to their activity. Now, near the end of its existence, this papilla will degenerate or be broken off near its base.

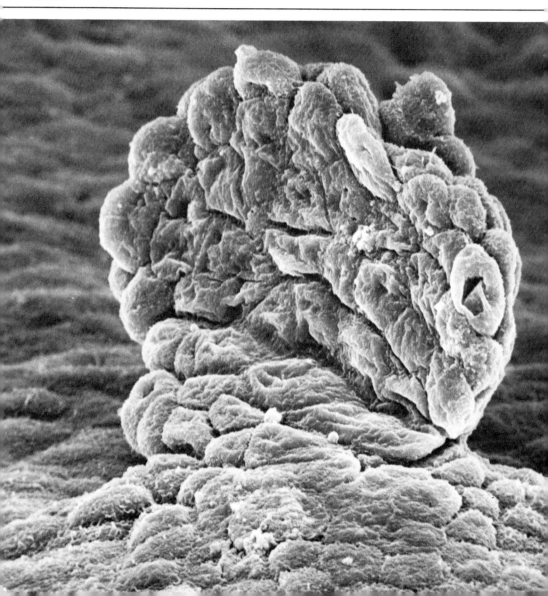

Summation

Our task here is to set down at least a minimal set of rules that may govern cell and tissue development. Any reader who has progressed through the whole of this book will recognize, first, that diversity is a primary characteristic of organ development. No two cell lineages seem to pass through identical histories, even though the final organs—say, salivary gland and pancreas—bear striking resemblance to each other. In part, this may be a consequence of the independent origins of various glands in evolution; thus, the starting point and the other cell and tissue types that are available would be pertinent to the way that a series of mutational events could lead to new developmental pathways and types of cells or organs.

254   Despite the unique history of each cell line, the following generalizations may help us gain a useful perspective.

1. *Cytoplasmic substances of the egg are a primary source of information for restrictive and expressive processes.*

Pole plasm, ascidian cortical materials, and gray crescent are models for what may be a general type of information that limits the ways that lineages of blastomeres can develop. Furthermore, analogous egg substances may trigger expressive events, such as gastrulation movements (recall pseudogastrulation) or enzyme synthesis (tyrosinase, cholinesterase).

2. *Position of egg cytoplasmic substances relative to each other and to the planes of cleavage is a primary source of information that affects much of development.*

The relative placement in a zygote of the cytoplasmic substances that affect restriction or expression is important for the placement of cell lineages (1) relative to each other, and (2) relative to sites of morphogenetic movements (which will rearrange lineages in space and so permit new interactions to occur).

3. *The changing position of cell lineages in time is a source of developmental information.*

Movement, changing mass, or relative growth of cell populations provide new information in an embryo. Juxtaposition of previously separate populations allows initiation of tissue interactions. The placement of a cell within the moving or expanding population affects its relationship to gradients of macromolecules (such as the ZPA factor in limbs) or to essential environmental parameters (such as oxygen).

4. *The cell surface is a primary site for processing of developmental information.*

The composition of the cell surface, in terms of receptors and, perhaps, transducers and primary effectors, is of crucial importance to cell and tissue interactions. Surface components derived from the zygote cortex, or possibly from other cell lineages due to interactions, could govern "competence" to respond to extrinsic signals. "Contact" with other cells, GAG, growth factors, hormones, or special features of the substratum could affect restrictive or expressive processes in a cell that has an appropriate detection and transduction system in its cell surface.

5. *Instructive tissue interactions can affect either restrictive or expressive processes.*

It has been demonstrated that one cell population can cause another to use genetic information it would not otherwise employ. However, the number of steps between cause and effect is so large that the causal sequences are yet unknown. Actions on the cell population (say, on GAG), on individual cell surfaces, or within responding cells are among the possible ways that chains of events could be initiated which culminate in restriction or determination. Direct "instruction" at the level of chromatin or genes is not required, though it is a possibility.

Rare cases of altered morphogenesis show that information from tissue interactions can govern the shape and assembly of epithelial cell populations. Similarly, some mature tissues retain a limited flexibility in differentiation, so that, depending on type of nerve, hormone, or other agent present, one developmental pathway or another may be followed by cell progeny. (If we assume that the restriction process stops in such epithelia at a point where a very few alternative differentiative pathways can be employed, then these cases do not do violence to the basic scheme for determination, differentiation, and tissue interactions.) These instances of "instruction" of expressive processes also do not demand primary action at the level of chromatin or genes.

6. *Permissive tissue interactions may be involved in any aspect of normal cell development, but do not govern the restriction-determination process.*

Permissive agents, which can include such things as intraocular pressure, hyaluronic acid, and estrogen, may be essential for morphogenesis, cellular differentiation, or even production of specific mRNAs or proteins. However, they are not, as far as we know, involved in setting up the *heritable* condition of determination.

7. *Attachment to or contact with a solid substratum is a basic permissive requirement for many types of normal cell behavior in development.*

The basal lamina, with its collagen and GAG, provides a surface that both stabilizes the shape of epithelial populations that are undergoing morphogenesis and provides anchorage for mitotic cells in epithelia. Bundles of collagen fibers, masses of GAG, or aggregations of cells may provide surfaces that influence cell locomotion, mitosis, or other cell activities in development.

256 8. *Major regulatory transitions in development tend to occur in cell populations carrying out mitotic cycles.*

All embryonic tissue interactions that result in restriction, in determination, and probably in the major initial transition to the differentiated state involve populations of cells that are carrying out mitosis. Unique events that involve nucleic acids may be able to occur only when the chromatin, the cell surface, or other parts of the cell are in the state that is characteristic of cells that are carrying out mitosis. In contrast, modulation in protein or nucleic acid syntheses, changes in distribution of molecules and structures, and many other processes related to the ongoing physiology of cells can occur in differentiated cells that are postmitotic (i.e., pancreas exocrine proteins in response to diet; skeletal muscle in response to nerve; etc.) Note: this generalization is the one most subject to argument, but also the one most open to experimental tests by means of currently available methodology.

9. *Tissue interactions occur in postembryonic organisms as sources of integration and of stability for cell phenotype (form and function).*

Products of neurons, hormones of the endocrine system, chalones, and growth factors are essential components of the intercellular environment that both maintains cell phenotype and controls rates of function. To the extent that the spectrum of required factors for a given cell is present constantly, the phenotype will be stable, and the cell will function appropriately from the organism's point of view. In a sense, it is superfluous that the determined state of such a cell is heritable. However, in emergencies, for example, wounding, or interruption of supply in environmental factors, the stability of the determined state provides a safety factor that insures that cells of the correct type will appear again once the intercellular milieu returns to normal. An alternative, of course, might be to "develop" such cells again by means of the tissue interactions and other processes that are employed in the embryo; difficulties with that scheme are obvious.

These "rules," in combination with the *concepts* listed at the end of each chapter, provide a framework for thought and planning. By paying increased attention to the precise way that cells and tissues respond to extrinsic information, we will be better able to understand the mechanisms by which tissue interactions help to insure coordinated development of diverse cell populations in animal embryos.

Glossary

acetylcholine A derivative of choline that may be a neurotransmitter that is liberated at endings of spinal-cord and parasympathetic motor nerves.

acinus (acini) A cluster of epithelial cells surrounding a cavity, or lumen, into which characteristic secretory products are passed; for example, acini are found in pancreas or salivary glands.

actinomycin D An antibiotic that binds preferentially to deoxyguanosine residues in DNA and thereby interferes with DNA-dependent RNA polymerase and, to a lesser extent, with DNA polymerase.

allantois An extraembryonic membrane of reptilian and avian embryos that serves primarily as a site of waste storage (e.g., of uric acid).

antigen A molecule or molecular complex that can elicit an immune response. Either a cellular reaction or a synthesis of circulating antibody may result from recognition of an antigen.

ascidian An organism in one of the groups of the urochordates (a primitive type of chordate). Ascidian tadpoles have tails that contain a notochord, striated muscle cells, and a dorsally situated nerve cord.

axolotl A type of salamander (uro-dele), frequently of large size and generally found in high mountain lakes and streams in Mexico and in the western United States.

barb, barbules Barbs are the individual strands of keratinized cells that make up the vane of a feather. Barbules are branches off the barbs, which, because of hooklets on their ends, hold the barbs together.

basal lamina A sheet of collagen meshwork and GAG that is intimately associated with the basal surface of epithelial cells.

blastema A mass of mitotically active cells, present at sites of regeneration, that gives rise to the regenerated structures.

blastomere An early embryonic cell that arises by the cleavage process.

5-bromodeoxyuridine A structural analogue of thymidine, in which a bromine atom takes the place of a methyl group. DNA polymerase cannot distinguish the two compounds; so bromodeoxyuridine is built into newly synthesized strands of DNA in place of thymidine.

chalone A putative cell-type specific inhibitor of mitosis.

258 **chromatin** The substance of chromosomes; it is composed of about one-third each of DNA, histone, and nonhistone protein.

ciliary ganglion A ganglion of the parasympathetic system, situated near the eyeball, and containing the cell bodies of motor cells that control the iris and the smooth muscle in the walls of the eye's blood vessels.

cleavage Mitotic cell divisons that follow fertilization and are characterized by an absence of net cell growth.

clone (cloning) A population of cells descended from a single cell.

cocoonase An enzyme manufactured by pupae of various insects that digests the cocoon and so permits emergence of the adult.

colcemid (colchicine) Alkaloids that bind mainly to tubulin, the building block of microtubules; as a result, these alkaloids interfere with microtubule-dependent processes.

collagen A class of proteins found in intercellular spaces and characterized by presence of proline, hydroxyproline, and hydroxylysine; individual collagen molecules are of highly elongated shape and can polymerize into various macromolecular aggregates.

compensatory hypertrophy A process in which part of an organ or organ system expands in size and functional capacity in order to compensate for lost or malfunctional tissue of the same type.

competence Ability of an embryonic cell population to respond to other tissues or agents by carrying out a developmental process.

cornea The outermost, transparent portion of the eyeball. Derived from head ectoderm and mesoderm.

cortisone A steroid hormone, derived from cholesterol, that is produced in the adrenal cortex. Among its many actions is stimulation of connective tissue to carry out repair processes.

cyclic nucleotide (cyclic AMP) A nucleotide (base-sugar-phosphate) in which phosphate is linked covalently to the hydroxyl groups on two carbon atoms, thereby forming a special cyclical portion of the molecule.

dermis Mesodermal tissue of skin.

determination Commitment of a cell to a specific mode of differentiation.

differentiation Process whereby a cell assumes a specific morphological and functional condition.

ecdysone A steroid hormone in insects that stimulates a variety of synthetic processes and causes molting. Ecdysones are also found in other arthropods and in some plants.

ectoderm The outer layer of animal embryos after gastrulation is complete.

egg (oocyte, ovum, zygote) Nomenclature varies. According to some, an oocyte is a fully grown female gamete (reproductive cell); an egg is a mature gamete that is in a fertilizable condition after ovulation; and a zygote is a fertilized egg.

electron microscopy The term formerly referred to "transmission" electron microscopy, in which electrons penetrate a relatively thin object to varying degrees, and so form an image of the density in the object. *See also,* **scanning electron microscopy.**

endoderm The innermost layer of animal embryos.

epidermis Outer, ectodermal tissue of skin.

erythropoietin A small, protein hormone that is produced by certain kidney cells when oxygen tensions or numbers of red blood cells are low. It stimulates red blood cell development.

estrogen A steroid hormone that is derived from cholesterol and produced by special cells of the ovary. It is the major female sex hormone.

5-fluorodeoxyuridine A structural analogue of thymidine that binds to the enzyme thymidylate synthetase, the enzyme responsible for the manufacture of thymine, a base in DNA. Lacking sufficient thymine, cells cannot divide.

follicle cell   A cell in the ovary of some animals that participates in maturation of oocytes.

ganglioside   Very large molecules containing a complex carbohydrate "head" group, glycerol, and long-chain "tails" composed of CH_2 groups. Associated with cell surfaces.

gastrulation   The gross rearrangement of cell populations (germ layers) that marks the completion of cleavage stages.

germ cells (primordial)   The stem cells in ovaries and testes that give rise to ova or sperm.

gizzard   The thick-walled, muscular, posterior portion of the avian stomach.

glycogen   A storage polysaccharide of animal cells in which D-glucose is linked in complex, branched chains. Frequently present in the form of several kinds of granules.

glycolytic pathways (glycolysis) Pathways for the anaerobic fermentation of glucose, and from which net ATP is generated.

glycoprotein   A protein- and sugar-containing macromolecule in which the quantity of protein exceeds the amount of sugar present (*see also*, **proteoglycan**). Characteristic sugars are linked to side chains of the amino acids of the protein.

glycosaminoglycans (GAG)   High molecular weight polymers of uronic acids and acetylated amino sugars; they may be sulfated.

gray crescent   A specialized portion of zygote cortex in amphibians and lungfish that corresponds to the site where gastrulation movements start, and that is associated with prospective axial mesoderm blastomeres.

histone   Small proteins with high contents of basic amino acids (arginine, lysine) that are associated with DNA in chromatin.

hyaluronic acid   A large, linear polymer with repeating units of D-glucuronic acid and N-acetylglucosamine. Found in basal laminae and the intercellular spaces of connective tissues.

hybridization, chemical   A process in which RNA or DNA molecules are caused to associate specifically with other RNA or DNA molecules that have complementary base sequences.

hydroxyurea   A derivative of urea that interrupts the DNA-synthesis *(S)* phase of the cell cycle.

imaginal disc   A small cluster of cells found in insect larvae that gives rise to a specific adult structure during metamorphosis; thus, there are discs for eye, limb, antenna, and so on.

insulin   A protein hormone produced by B-cells of the Islets of Langerhans in the pancreas. It has many effects, including ones on blood glucose levels. It may act by inhibiting adenyl cyclase activity and thereby lowering the amount of cyclic AMP in cells.

integral protein   Singer's term for cell surface proteins that can only be removed from the plasma membrane lipid bilayer through the use of strong, hydrophobic bond-breaking agents. Because of their hydrophobic surface properties, integral proteins are believed to be situated deep within the lipid bilayer.

iris   The rim of opaque tissue surrounding the pupil of the eye.

junctions (gap, septate, desmosomal, etc.)   Specialized areas of contact between cells. Unique intramembranous particle distributions, close apposition of plasma membranes, and nearby cytoplasmic specializations characterize the various types of junctions.

juvenile hormone   A long chain hydrocarbon of complex structure.

260 This insect hormone prevents cells from maturing during larval or pupal stages.

keratin A protein complex with high sulfur content found in differentiated epidermal, hair, feather, and scale cells.

kidney The vertebrate excretory organ that eliminates blood-borne wastes.

lac operon The region of genetic material on the *E. coli* chromosome that controls synthesis of β-galactosidase. Repressor protein binds to the regulatory gene which controls synthesis of the mRNA that encodes for the enzyme.

larva A general term for the immature stage of a species that follows the embryonic period (e.g., a tadpole, caterpillar, etc.)

lens The spherical body, located just behind the pupil of the eye, which bends light rays so that an image is focused on the neural retina.

limb bones In proximodistal sequence: in the fore limb, humerus, radius and ulna, carpels, metacarpels, phalanges; in the hind limb, femur, tibia and fibula, tarsals, metatarsals, phalanges.

medullary plate The area of ectoderm that thickens, undergoes morphogenesis, and gives rise to much of the vertebrate nervous system.

meiosis Reduction divisions that produce haploid sex cells (gametes) from diploid germ line cells.

melanin A black or brown pigment composed of oxidized derivatives of the amino acid tyrosine and found in cells in the form of dense granules.

mesenchyme A mass of embryonic connective tissue cells; derived from mesoderm.

mesoderm The middle layer(s) of the animal embryo.

microfilament Cytoplasmic filaments, composed of actin, that may play a structural or contractile role in nonmuscle cells.

micron A length equal to 10,000 Ångstrom units or to 0.001 mm.

microtubule Elongate, rodlike structures composed of tubulin and believed to play a skeletal role in some cells.

mitosis Cell division in which duplication and separation of chromosomes and division of the cytoplasm occurs.

morphogenesis The process in which cells or cell populations undergo changes in shape or position incident to development.

motor end plate The specialized junction of a motor nerve ending and a muscle cell. Site of acetyl cholinesterase, sensitivity to neurotransmitter, and various specialized structures.

mRNA RNA molecules that act as specific templates for synthesis of proteins.

mucus Secretion of mucous glands or epithelia. Composed of protein and carbohydrate and serving as a lubricant and protectant.

myoblast A prospective skeletal muscle cell. Myoblasts fuse to form myotubes, which differentiate into skeletal muscle cells.

neural tube The embryonic spinal cord of the vertebrate central nervous system.

neuroblast An immature nerve cell that still has the capacities for mitosis and locomotion.

notochord A rigid, cellular rod located beneath the embryonic spinal cord; the functional forerunner of the vertebral centra.

optic vesicle A hollow derivative of the medullary plate, the walls of which are continuous with the brain walls, and which gives rise to the pigmented and neural retinae.

ovalbumin The major egg white protein (MW 43,000 daltons) of reptiles and birds.

papaverine   An alkaloid that prevents smooth muscle contraction from starting or causes relaxation of contracted muscle cells. It may act by affecting availability of calcium ions to cells.

parabiosis   Technique of linking two individuals so that common flow of the blood vascular system exists. This permits radical operations on one of the pair.

parasympathetic ganglia   Ganglia of the autonomic nervous system of vertebrates that are the site of neurons which release acetylcholine, to decrease blood pressure, close the iris, inhibit salivary secretion, and so forth.

particles, intramembranous   Particles situated between the inner and outer lipid layers of the plasma membrane. Believed to be proteins.

peripheral protein   Singer's term for cell surface proteins that can be displaced from the plasma membrane by use of mild conditions or agents (ionic strength, pH, etc.). Such proteins are generally equivalent to "soluble" proteins in aqueous solutions.

ploidy   Refers to the number of chromosome sets present in a cell (haploidy, diploidy, and so on).

pole plasm   Material concentrated near one end of an insect egg that is required for germ cell preservation and development.

polydactyly   A condition in vertebrates in which extra digits occur on a limb.

polymerase (RNA, DNA)   Any of the enzyme activities that are essential for the synthesis of RNA or DNA molecules that are complementary to nucleic acid templates present in the reaction mixture (or in nuclei).

polytene chromosome   Giant, interphase chromosomes of certain insects, in which the basic DNA double helix backbone and associated proteins have been replicated up to several thousand fold. The multiple copies are aligned precisely side by side so that genetic loci are in register.

progesterone   A steroid hormone derived from cholesterol and produced by placentae or special ovarian cells.

proteoglycan   A protein- and sugar-containing macromolecule in which the quantity of sugar far exceeds the quantity of protein (*see also*, glycoprotein). Characteristic sugars are present in these components of the extracellular spaces of tissues.

proventriculus   The anterior portion of the avian stomach. It is the site of many glands and of digestive enzyme secretion.

puffs   The magic dragons of developmental genetics are in reality regions of insect polytene chromosomes in which the many copies of the DNA-protein backbone are spread apart from each other. Intense RNA synthesis is found in puff regions.

pupa   A stage in the life cycle of insects that follows the larval period and precedes adulthood. Frequently the pupa is encased in a cocoon and is quiescent behaviorally as larval tissues break down and adult ones develop.

rachis   Shaft or "backbone" of a feather.

salivary glands   Submandibular, sublingual, and parotid glands; responsible for production and secretion of salivary components; also, storage sites for NGF and EGF.

sarcomere   The contractile units of skeletal muscle cells; contain actin, myosin, and numerous other specific proteins arranged in a characteristic manner.

scanning electron microscopy   A procedure in which electrons bombard the dried surface of an object and generate secondary electrons which can be detected and used to form an image of the object.

Schwann sheath cell   Cell derived from embryonic neural crest. Schwann cells invest neurites of the peripheral nervous system with myelin, a form of insulation.

262 **sclera** The connective tissue sheath surrounding and protecting the eyeball. Scleral bones are a ring of dermal membrane bone situated around the rim of the avian cornea.

serine An amino acid possessing an hydroxyl group on the side chain. Serine is usually "nonessential" for cultured mammalian cells, meaning that it can be manufactured by such cells from other organic substances.

somatomedin A small protein that is probably released from liver in response to growth hormone. It stimulates a variety of synthetic processes in several cell types but especially bones.

somite Masses of mesodermal cells situated lateral to the neural tube in vertebrate embryos. Cells from somites form dermis, cartilage, bone, muscle, and other tissues.

spinal cord In vertebrates, the elongate, hollow portion of the central nervous system that extends posteriorly from the brain down the axis of the body.

spinal ganglion Also called sensory ganglion and dorsal root ganglion. Ganglia located lateral to each side of the spinal cord in vertebrates; the sites of sensory nerve cell bodies.

stem cell "Germinal" cells of an epithelium or of certain mesenchymal populations that divide at controlled rates in adult life, and give rise to differentiating progeny cells (e.g., in taste buds, the blood lines, epidermis).

steroid A type of lipid that cannot be rendered water soluble with alkali; built of a complex four-ringed structure (as in cholesterol), and most often possessing hydroxyl groups. Steroids are an important component of cells from all higher organisms, and may have important regulatory activities (as, sex hormones).

sympathetic ganglia Ganglia of the autonomic nervous system of vertebrates that are the site of neurons which release norepinephrin (noradrenalin) to stimulate various activities, raise blood pressure, and so forth.

taste bud Aggregations of taste receptor cells on the tongue (or head of fishes), with sensitivity to sweet, sour, salt, and bitter modalities. The receptor cells depolarize, and trigger impulses directed toward the brain over neurites of certain cranial nerves.

tertiary protein Proteins unique to single types of differentiated cells; e.g., insulin for B-cells, hemoglobin for erythrocytes.

testosterone A steroid that is a male sex hormone.

2-thiouracil A sulfur-containing derivative of the base uracil. It inhibits production or secretion of the thyroid hormone(s).

time lapse cinematography Motion pictures of living systems in which photographic exposures (the frames on the film) are made at relatively widely spaced intervals; when these are projected at faster speeds, "movement" of the systems can be observed and studied.

transcription The process whereby a sequence of DNA is copied into a complementary sequence of RNA, as in mRNA production.

translation The process wherein a specific mRNA is used as a template for assembly of amino acids into a polypeptide.

trophic factor A hypothetical agent required for survival, function, or development of a cell, particularly in the nervous system.

tubular gland(s) Clusters of secretory cells in the wall of the avian oviduct. Site of synthesis and secretion of various egg white proteins.

tyrosinase An intracellular enzyme that acts at an early step in the production of melanin from the amino acid tyrosine.

ureteric bud An evagination of the urinary duct system in embryos of land vertebrates, that branches and gives rise to the pelvic, collecting duct portion of the adult metanephric kidney.

uronic acid A sugar in which the carbon atom with the primary hydroxyl group is oxidized to form a carboxyl group, such as D-glucuronic acid from D-glucose. Uronic acid is a component of all GAGs.

zymogen granule Large storage granules in secretory cells that contain an inactive precursor form of an enzyme or protein (the zymogen).

Index